JOHN GLOAG ON INDUSTRIAL DESIGN

Volume 1

INDUSTRIAL ART EXPLAINED

INDUSTRIAL ART
EXPLAINED

JOHN GLOAG

Routledge
Taylor & Francis Group

LONDON AND NEW YORK

First published in 1934 by George Allen & Unwin Ltd.

This edition first published in 2023
by Routledge
4 Park Square, Milton Park, Abingdon, Oxon OX14 4RN

and by Routledge
605 Third Avenue, New York, NY 10158

Routledge is an imprint of the Taylor & Francis Group, an informa business

British Library Cataloguing in Publication Data
A catalogue record for this book is available from the British Library

ISBN: 978-1-032-36309-7 (Set)
ISBN: 978-1-032-36526-8 (Volume 1) (hbk)
ISBN: 978-1-032-36531-2 (Volume 1) (pbk)
ISBN: 978-1-003-33250-3 (Volume 1) (ebk)

DOI: 10.1201/9781003332503

Publisher's Note
The publisher has gone to great lengths to ensure the quality of this reprint but points out that some imperfections in the original copies may be apparent.

Disclaimer
The publisher has made every effort to trace copyright holders and would welcome correspondence from those they have been unable to trace.

INDUSTRIAL ART
EXPLAINED

by
JOHN GLOAG

Illustrated by
NORMAN HOWARD
and from Photographs

Published for
THE SCIENTIFIC BOOK CLUB
121 CHARING CROSS ROAD
LONDON, W.C.2
by
George Allen & Unwin Ltd.

FIRST PUBLISHED IN 1934
THIS EDITION PUBLISHED IN 1946

Printed by Bradford & Dickens, London, W.C.1

DEDICATED TO
FRANK PICK

ACKNOWLEDGEMENTS

I WISH to acknowledge the courtesy of Mr. Thomas Falconer, F.R.I.B.A., in permitting me to reproduce the plates of the eighteenth-century factories in the Stroud Valley, and of Messrs. Pilkington Brothers, Limited, in providing the illustration of the old casting hall at Ravenhead, St. Helens. To Mr. Frank L. Heathorn, and to His Master's Voice Gramophone Company I am indebted for putting at my disposal innumerable illustrations and abundant information regarding the evolution of gramophone forms. To the manufacturers and designers who have allowed me to reproduce examples of their work, and to Mr. Norman Howard whose skill, patience, and research have made the illustrations in the text so lively and informative, I am duly grateful: without their help it would have been impossible for this book to suggest that design in industry is a living force.

JOHN GLOAG

CONTENTS

LIST OF LINE ILLUSTRATIONS IN
THE TEXT

LIST OF LINE ILLUSTRATIONS IN THE TEXT

LIST OF PLATES

CLASSIFIED LIST OF ILLUSTRATIONS

HERE both line illustrations in the text and the plates have been classified under the following headings:

INDUSTRIAL DESIGN

MACHINE DESIGN

INDUSTRIAL ARCHITECTURE

MACHINE ARCHITECTURE

PRE-INDUSTRIAL DESIGN

INDUSTRIAL DESIGN

INDUSTRIAL ART EXPLAINED

MACHINE DESIGN

CLASSIFIED LIST OF ILLUSTRATIONS

INTRODUCTION

THE CASE FOR AN ACADEMY OF DESIGN

R EADERS who wish to skip this introduction
will only be missing the sermon. The real business of this book begins with Chapter I. But even
before the sermon begins, there are some things which
should be explained. For instance, there are several
long quotations from a nineteenth-century author of
guide-books named Samuel Sidney. His was an ironi-
cally modern voice and he observed and criticized
the world about him, and particularly its artistic
endeavours, with an outlook not unlike that of
practitioners of the modern movement in the nine-
teen-thirties. The writer is indebted to Mr. C. C. J.
Simmonds for introducing him to that rare and enter-
taining little volume called *Railway Rides* in which
Samuel Sidney set down his thoughts with a mordant
pen. Samuel Sidney's observations illustrate the growth
of muddled thinking about the whole question of
industrial design in the industrial century, and in his
pages one constantly gets the impression that he realized
that something was drastically wrong but was unable
to diagnose the cause although confronted by an
array of depressing symptoms.

There are two long quotations from previous books
by the writer which have been made because they
deal specifically with certain phases of design in the
nineteenth century and with the development of the

artist-craftsman in the first two decades of the twentieth century.

This book does not set out to be a detailed and academic history of design in industry, or an economic-cum-artistic survey of the tremendous field covered by the term industrial art. A capable outline for an economic survey already exists. It is concentrated in one small but potent volume by Sir Hubert Llewellyn Smith and is called *The Economic Laws of Art Production*.

Only a few branches of industry have been chosen for illustrating points about design. Any attempt to examine the problems of design connected with all branches of modern industry would demand, not one small book, but several large ones.

Readers who survive until the end of the last chapter may conclude that the whole book is directly and indirectly concerned with putting a case for the employment of architects in far more departments of design than building. That is an inevitable conclusion, because the architect is a trained designer, and design in industry can only be improved by the confident employment by manufacturers of competent designers. "To cultivate an eye for architecture, whether in the abstract or for its human revelation, is to cultivate an eye for all design," writes Robert Byron in *The Appreciation of Architecture*; "in other words, to cultivate a technical and absorbing interest in the whole works of man. For what are the visible works of man but the fruit of design laboriously committed to paper before being realized beneath the tools of the craftsman or the stamp and mould of machinery? The tie you wear, the print you read, the pen you hold, the lamp that lights the room, the kettle on the stove, the carpet on the floor, the paper on the wall, the posters on the hoardings, the grocer's van, the public convenience—every object bears the stamp of some unwitting or some would-be artistry, and can be assessed or dismissed in terms of artistry."

And that is the real text of the sermon. Many people have delivered it before; and perhaps the ablest preacher whose words converted innumerable manufacturers was the late Sir Lawrence Weaver, the best friend designers have had this century. It has been given in the past and is being given again here because in many branches of British industry there is a deep-rooted belief that good design does not pay and could never be made to pay. When any manufacturer displays signs of imagination, one suspects that finance, as represented by that queer uncreative machine, the accountant, intervenes repressively. The accountant speaks and thinks of "tangible assets," and is often incapable of imagining that anything exists that cannot be recorded upon a balance sheet. This belief is partly responsible for nearly every country in Europe being far ahead of England in those sections of industry that depend for their continued health upon a working partnership with design. How long manufacturers who underrate the importance of this partnership can avoid financial as well as artistic bankruptcy depends upon the pace of education at home; but the shrinking of their business owing to the loss of foreign markets

25

will come long before the final oblivion that awaits the unprogressive concern.

In civilized countries it seems obvious that manufacturers and designers should work together for their mutual benefit. Most European countries have organizations to facilitate such contacts, and the directors of industry support those organizations and thereby give them the financial stability that enables them to render practical service. Attempts have been made to run such organizations in England; and the history of those attempts and the success and character of the societies and associations that have been brought into being are perhaps the most severe comments upon commercial perception that the record of an industrial nation could afford. The two or three guineas per annum that a minute number of firms contributes to these associations are too often regarded as doles to bolster up "ideals" and are liable to be lopped off budgeted expenditure any year.

The two most important organizations that could act as liaisons between art and industry and are potentially of immense value for this purpose are the Royal Society of Arts, and the Design and Industries Association.

The Society of Arts was founded in 1754, was incorporated by Royal Charter in 1847, and in 1908 was allowed to add "Royal" to its title. The object of its founders was "the Encouragement of the Arts, Manufactures, and Commerce of the Country" and their encouragement was planned to take the form of rewards for inventions and improvements in various departments of industry and applied science and art, and for work of merit in the Fine Arts. Agriculture, chemistry and branches of engineerings came within the scope of the Society's terms of patronage; and courses of improving and interesting lectures became part of the yearly programme carried out in the spacious and beautiful building in John Street, Adelphi, that houses the Society. Since 1854 examinations have been held three times a year under the auspices of the Society in subjects that include "the principal elements of Commercial Education and Modern Languages."[1] Some years ago Mr. Stanley Baldwin blessed a further extension of the Society's activities, and incidentally disclosed his ideas of the function of a body dedicated to the "Encouragement of the Arts, Manufactures, and Commerce of the Country," by inaugurating a fund for the preservation of Ancient Cottages. It is to the credit of the Society that it was the first body to identify the existence of industrial art. In Chapter II some of its early attempts to make Victorian manufacturers realize their responsibilities about design are recorded. Unfortunately the Society suffers from a diversity of intention, and although it has during its long lifetime created and maintained an organization that handles its departments in a highly efficient manner, it cannot encourage "the Arts, Manufactures, and Commerce of the Country" unless it is prepared to lead industry.

The Design and Industries Association could never have been formed in a really civilized country, because

[1] Quoted from the *Proceedings of the Society*.

27

no civilized country would have permitted industry to get into a condition in which its products were so fantastically inept that manufacturers had to be reminded that fitness for purpose was a basic principle of design in everything—tea-pots, tables, houses, hot-water bottles and whisky decanters. It was with the object of propagating the elementary common sense represented by the phrase "fitness for purpose" that the D.I.A. was founded in 1915. It has limped along, crippled by lack of funds, ever since; and its propaganda has not been without influence, although that influence has been limited by lack of support from the very section of the community that the Association exists to serve.

If organizations that have common aims could merge, perhaps to their mutual invigoration, a concentrated effort to improve English standards of industrial design might be made. An obvious merger must come to the mind of anyone who gives thought to the needs of industrial art, namely: The Royal Society of Arts and The Design and Industries Association. Such bodies in association could begin to represent design and designers powerfully, and commercially. It is only by commercial representation that they could convince manufacturers that they were worth consideration. The independent bodies that attempt to make industry attend to its business with art are usually ignored or half-contemptuously humoured by manufacturers, because it is sometimes difficult for manufacturers to imagine that people who are giving time and energy to propaganda and education in a purely disinterested spirit can possibly be serious.

If therefore one authoritative Society of Design was formed from existing societies, and represented the designer, becoming a sales organization for his work, it would have more respect from the business community which would at last begin to understand that design could become a "tangible asset." When people are told they must pay for a thing they set a value upon it which may, or may not, correspond with the price asked : they may not buy it, but they will never dismiss it as valueless, when once a price has been asked. Once design and designers are given commercial standing it should be easier to attract industry to a partnership with art.

At the moment, if a firm of manufacturers wants to make some experimental designs with a new material that can replace some accustomed material in the making of furniture, for example, how can its directors get in touch with designers best qualified to serve their particular needs? It is pure luck, nine times out of ten, if they employ the right man or woman to advise them : generally they get the wrong one, and the resulting failure discredits design for five or ten years in that particular business. If a Society or Academy of Design existed it could give immediate and practical help to such a firm. It would make itself known to manufacturers through contact with the various Chambers of Trade and Commerce up and down England. Its story could be bluntly condensed into a variant of a famous advertisement slogan : "You want the best art, We have it !"

The Society would know, what Government Departments seldom know, who were the most intelligent and
able designers for any type of work that could be called
for on those occasions when Britain has to express herself artistically to an international audience. Generally when such occasions occur, Europe titters at
the results, and America says: "Now isn't that just
the cutest bit of old-world design!" Instead of making
ourselves ridiculous in the markets of the world with
Tudor exhibition pavilions, and imitations of things
produced by craftsmen of the seventeenth and eighteenth centuries, we might be able to prove that the
art of the country that produced Wren and Chippendale and Wedgwood still lived and breathed.

This dream Academy of Design would organize
exhibitions of contemporary work in every manufacturing centre in England: it would include Continental exhibits, not for imitation, but to make manufacturers realize that there was a modern movement
in every branch of design, and to illustrate its different
national manifestations. A Government subsidy to support and extend its work would have far more effect
upon standards of taste in England, and would do far
more to elevate artistic appreciation, than the artificial respiration of Grand Opera. These suggestions
about the amalgamation of the various organizations
concerned with design and the formation of an Academy
of Design from their merged activities were originally
made in *The Architectural Review* in February 1931 by
the writer. The Academy of Design is still a dream;
but the Royal Society of Arts by collaborating with the

Royal Academy in organizing an Exhibition of Industrial Art at Burlington House has courageously resumed its responsibility for the improvement of design in "manufactures and commerce."

Responsibility for education in design is wastefully diffused. The possession of an orderly mind ensures the maximum of irritation in the commercial machine age. But some comfort can be derived by recognizing the individual excellence of many existing bodies, such as the Royal Society of Arts, the Royal Institute of British Architects, the Design and Industries Association, the Council for the Preservation of Rural England, the Society of Industrial Artists, etc., and hoping that one day their not unrelated activities and responsibilities may be co-ordinated, and their energy pooled for the benefit of a G.H.Q. of design—the remote, improbable, Utopian, but excessively desirable Academy of Design. One day there may be a Ministry of Planning, but that is perhaps too far ahead; it may only come years after the ashes of the bright young designers of to-day have been reverently enclosed not in urns, but in gleaming cylinders of chromium-plated steel, or scattered from the topmost balconies of the residential towers that, in the far future, may replace slums and ribbon development.

INDUSTRIAL ART EXPLAINED

CHAPTER I

INDUSTRIAL ART EXPLAINED

THE growth of mechanized industry during the last hundred and fifty years has affected our surroundings and the articles we use every day. It is the purpose of this book to show how design has affected:

1. The accommodation of industry,
2. The products of industry, and,
3. The distribution of those products to the public.

The term *industrial architecture* will be used for the first category; *industrial art* will apply to the second, and *commercial art* will cover the third. Reference will also be made to the nature and development of what may be called *machine design*.

The way in which machines are housed and arranged, the way in which active labour is provided with shelter, ventilation, light, heat and apparatus in a factory, and the meeting of these mechanical and industrial needs, have influenced modern architectural design. Industrial architecture appears when the most capable technical solution of the problems of housing plant and organizing production have been made.

When the builders of factories in the last half of the nineteenth century thought about architecture they thought of it as a disguise, and they tried to make their factories look externally like Venetian Palaces, or something equally unsuitable. The architect was, in the eyes of the factory-owner, somebody who covered up the work of the engineer—a mere provider of stage scenery—and in many places Victorian factories still hide their functional features behind the false whiskers of some drawing-board "style," something Florentine or Gothic.

Machinery came into a world that distrusted and disliked it. Very naturally manufacturers took to disguising not only their factories but the things made in them. Who in those early days of industrial production could have ventured to make articles designed expressly for machine production, and possessing in consequence an entirely new character? The market for machine-made goods had to be wooed by cheapness and by mimicry; the manufacturer had to say: "These things I am making are dirt cheap, and look! they are just as good as the old hand-made things." Only in entirely new industrial or engineering activities could anything be designed freely and without ideas being conditioned by the form of some prototype. For instance with railways, locomotives were freely designed. They were not at first built for high speed, because although the early locomotive engineers were daring and ambitious their minds were conditioned, as indeed was the public mind of that time, by the speed of the horse. Twelve miles an hour was considered dangerous. Twenty miles an hour was tempting providence. Perhaps it was with those enormously high chimney

34

stacks tied back with stays to the sides of the boiler; but directly the idea of ·high speeds was generally accepted, locomotive design changed rapidly. In a quarter of a century it changed from this:

FIG. 1—*The Rocket, 1829*

to this:

FIG. 2—*The next stage of locomotive design: The Lion*

Before 1900 it had achieved these forms:

FIG. 3—*The sleek, trim lines of the modern locomotive are foreshadowed in this Great Western Railway type*

FIG. 4—*Late nineteenth century. (Great Northern type)*

to-day it is like this:

FIG. 5—*The Princess Royal. (London Midland and Scottish Pacific type, 1934)*

and to-morrow this may be the common form,

FIG. 6—*The streamlined train with locomotive and coaches as continuous flexible units*

with locomotive, and coaches as flexible, streamlined units.

The locomotive is an example of machine design. The engineer was under no imagined obligation to make it anything but a compact, economical solution of a problem. And an orderly and trim solution the English engineers produced; in every aspect superior to the untidy monsters that hauled Continental and American trains, with complex apparatus exposed all over their external surfaces. The design of railway coaches was hampered by the memory of an earlier machine: the stage coach. Also the habit of putting private carriages on to flat trucks for transport by rail-

FIG. 7—*The private carriage of the early nineteenth century*

way maintained the association of road vehicle forms with the development of railway coaches. The early first class carriages on English railways were two or

FIG. 8—*The stage coach of the early nineteenth century*

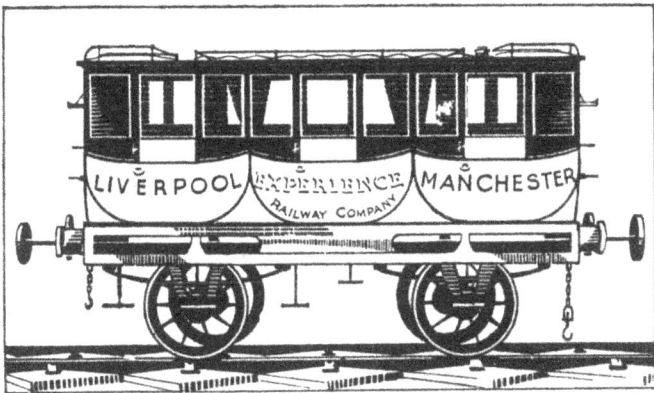

FIG. 9—*A first class railway coach, circa 1835*

three conjoined stage coaches. These forms were retained well into the present century, and they are the
basis of the compartment system, which persists to
this day, despite the world-wide influence of George

39

FIG. 10—*The corridor type of modern railway coach still bearing traces of its stage coach ancestry in the grouping of the compartments*

FIG. 11—*The Pullman car which was a complete break-away from the stage coach tradition of design*

Mortimer Pullman. Our only concession to mobility for passengers in England is the corridor, unsymmetrically disposed along one side of the row of old stage coach "insides." Our greatest achievement in designing for comfort is the sleeping berth.

The designing of rolling stock for the railways entailed different responsibilities from the manufacturing of commodities which had to be sold to the public. The railways had a monopoly; they were under no economic compulsion to cajole the public to travel by any artifice of form; they adapted a ready-made form, as the motor-car and motor-bus designers did at the beginning of the present century.

The machine production of furniture, for example, created an entirely different outlook upon design. Manufacturers felt compelled to imitate not the exact form of hand-made furniture, but the technique of hand-made furniture. The materials used for making furniture, the facilities afforded by the new mechanical methods, and the functional needs of such familiar articles as chairs, tables, sideboards and cabinets were never lucidly co-ordinated by a designer, so that a new technique of production could be evolved with its own characteristic forms. What may be called machine-craft never came alive in furniture manufacturing. Mechanical production was unacknowledged by the finished article in this and in other industries.

Industrial art was not recognized. A rich confusion of terms such as *decorative art, applied art* (the most unfortunate of all) and *ornamental art* prevented the emergence of the idea that from the power of machines and the nature of new materials there might arise, under the direction of designers, a new form of common art which would replace the old common art of England.

When design is free trom the influence of a proto-type and remains a purely functional problem of engi-neering, the beauty of mechanical fitness is achieved by an apt solution. It is a beauty different in kind from that created by the artist; it is a mathematical by-product: unsought: incidental. To imagine that beauty invariably résides in a piece of competent engineering and to assume that "fitness for purpose" is the golden rule for the creation of beauty is to chain creative ideas to abject utility. Industrial architecture has the beauty of mechanical fitness. Industrial art begins when the aesthetic judgment of a designer is employed to determine the character of a manufactured article.

The third section in which design is concerned with commerce and industry, can be described under the heading of commercial art. This has developed greatly during the last quarter of a century; parti-cularly since the war. It is concerned with the selling of goods and services to the consuming public. When the term "commercial art" is mentioned, most people think of posters; but posters represent only a relatively small part of the output of commercial art, for under this heading we must classify all illustrations in adver-tisements in the press, and also the design of tins and packets and wrappings and all printed matter distri-buted through the post and over counters which is intended to interest people in goods or services. The whole of the physical side of advertising is affected by commercial art; and upon this section the work of the printing industry impinges.

Under these three headings, Industrial Architecture, Industrial Art, and Commercial Art, we can examine not only the operation of design in industry, but the influence of industry upon the ideas of designers. The character of industrial building has already had an effect upon the architectural profession, and though this is deplored by some architects of the old school, it has been a healthy effect, for it has compelled an extensive recognition of the structural revolution that has taken place since steel and concrete have allowed buildings to be supported by an internal skeleton instead of·by external walls. As all revolutions for a time become the prey of extremists, there is a school of architectural design which has tried to foist upon the world the inhuman formula that "the house is a machine for living in." This has been responsible for much careful devising of intentional bleakness, so that a mechanical air may be suggested by an interior. A style which might perhaps be called mechanistic baroque has been cultivated. When this style is based upon imitation functionalism it is as stupid as imitation Jacobean. For instance, if window sills in a living-room or a kitchen are deliberately finished off at an angle of forty-five degrees to give an impression of sleek, dust-proof efficiency, imitation functionalism is obviously destroying ordinary convenience.

Mechanical forms are used as patterns for textiles and wallpapers; they provide inspiration for the mechanistic baroque style of decoration, but they do not represent a real industrial contribution to design. They represent the designer's appreciation not only of the external forms of mechanical apparatus, but of the character of the machine age. The designer's employment of such motives in the creation of patterns and decoration is an acknowledgement of a discovery which was made before the war by the Cubists. As Mr. R. H. Wilenski reminds us: "Gradually, all the world is beginning to realize that we live in an age of co-ordinated effort, of instantaneous world-wide communications, of moving photographs and aeroplanes, of skyscrapers, of steel and concrete factories, and scientific skill; that this age has a pattern of its own as all ages have had before it; and that the Cubist artists, who were ridiculed and reviled, were the first to recognize this pattern, to accept it, and to resolve to work out its development."[1]

But the commercial machine age is only beginning to be recognized as an age with a special character of its own. It is so full of perplexing and obstructive survivals; and of these the idea that it is unpractical to employ designers to improve the nature of industrial products has until recently been persistent. Even now, when manufacturers are prepared to try occasional experiments, there is a difficulty about the responsibility for patronage of design. But before discussing that problem of patronage, the history of Industrial Architecture and Industrial Art should be briefly examined.

[1] *A Miniature History of European Art*, chap. vii, p. 75.

CHAPTER II

A SHORT HISTORY OF DESIGN IN INDUSTRY

"The nineteenth century was the Age of Faith
in Fine Art. The results are before us."
Maxims for Revolutionists'
by GEORGE BERNARD SHAW

L OOKING back over the social and industrial
history of the last century and a half, it will be
seen that industry invaded an established form of
society without any plan being made for its accommodation in that society or without any thought being
given to its effect upon the face of the land and the lives
of the people. A rapidly growing population provided
factory fodder for industrial concerns which were everywhere expanding. At first industry grew up within the
existing social framework. It was accommodated in a
familiar and not unagreeable manner. Mills that were
built in the eighteenth and early nineteenth centuries
inherited some of the architectural graces of the great
age of English building, even though they were just
plain statements of need in brick, well lit and airy.
Some of the best examples of this early industrial
building are to be found to-day in the Stroud Valley.[1]
The architectural tradition of design was still strong
enough to control the form of any building that was
erected in England. It was only when machinery became more complex, and made special and unpre-

[1] See Plate I.

cedented demands for accommodation, that the form of factories and the whole character of industrial plant broke away from their existing architectural concept. The engineer with no architectural training had to provide shelter for the machines under his care.[1]

Technical ability in engineering was advancing so rapidly that even in the early factories and shops there was an inclination to accept makeshift arrangements for accommodation in order to allow for expansion, and to allow also for changes in the disposition and character of machinery. The account of the growth of the Soho Factory of Boulton and Watt at Birmingham in the late eighteenth century given by Mr. Erich Roll in *An Early Experiment in Industrial Organization* illustrates the reluctance of industrial pioneers to adopt any building programme which was not dependent upon the adaptation of existing buildings and the piecemeal addition of fresh shops. Factories it seemed were destined to grow untidily as mechanized industry developed.

By the eighteen-thirties all sorts of people were becoming conscious of a dirty disorder that was

---

[1] At first a stalwart architectural tradition guided him. See Plate II.

[2] "The Soho building, which was originally used for the manufacture of silver and plated goods, was in the form of the letter E, the wings of which were used as dwellings for managers and foremen. Two yards lay behind it, the first a story lower than the ground at the front, the second still lower. All the additional industries which arose at Soho were carried on in buildings erected in these yards." *An Early Experiment in Industrial Organization*, part ii, chap. i, p. 153.

afflicting towns and cities everywhere. The unplanned growth of industry took place at a time when the romantic movement, of which the Gothic Revival was one manifestation, was breaking down the idea of plan and architectural order in people's minds. Ruskin was defending architectural anarchy, and many pious and business-like architects were making spirited attempts to attain it. In the preceding age the design of nearly every object had been controlled directly or indirectly by architects. From the days of Charles II. to the death of George IV. the architect had influenced the design of everything from coach lamps to door knockers, and perhaps the greatest misfortune of the nineteenth century was the so-called "Battle of the Styles," with the protagonists of order and lucidity fighting with ever-diminishing strength the Gothic Revivalists who were determined to shape architecture by emotion and not by any principles of design. But even the ardent Gothicists were conscious of the bleakness and disorder created by industry. It was the wrong sort of disorder—it was not the fine, free, emotional muddle of romantic mediaevalism: it was a casual and godless exhibition of mechanical progress. Pugin in his book of contrasts which was published in 1836 (at his own expense) delineated two imaginary views of a town, in 1440 and in 1840. In the latter, ironworks and gasworks, huge warehouses, grim little chapels, lunatic asylums and so forth occupied the sites of the abbeys, churches, guild-halls and monasteries that had upraised their towers and spires in the mediaeval city. The whole book was a romantic pro-

47

test against industrialism, and Pugin as an architect realized that industry was largely undirected by design.

The architect had ceased to influence design apart from public and domestic building. He had little or no contact with industry. So factories were built in a shortsighted hand-to-mouth way, and the study of many industrial organizations to-day will show, save in exceptionally progressive instances, a patchwork of factories, square miles of industrial chaos often encircling a core of early nineteenth-century brick-built offices and shops which retain their architectural comeliness beneath the accumulated grime of a hundred-and-thirty-odd years. It is only to-day and only when industrial concerns are able to start with an unencumbered site that great industrial architecture is possible, of the kind created by Sir Owen Williams in this country and by Erich Mendelssohn in Germany. The factory built by Sir Owen Williams for Messrs. Boots at Nottingham and his extension to the London premises of Messrs. Sainsbury are outstanding examples of architecture which is produced by the scientific adjustment of modern building materials and constructional methods to the specific needs of factory organization. Modern industrial architecture of this type takes account of many things which would have been ignored in the nineteenth century. The buildings it produces are not wholly dependent upon concrete and steel and great expanses of glass for their light, airy and vigorous lines. They derive some of their character from the fact that they have been designed to create conditions which the early nineteenth-century manufacturer would have regarded as dangerously Utopian.

In spite of the inhumanitarians, the Manchester School individualists, factory conditions to-day transcend the dreams of any nineteenth-century workman. The really capable designer has come into the service of industry in the planning of factories, so that the needs of the different technicians can be co-ordinated and some lucid plan evolved for the accommodation of all the processes of production and all the plant required to operate a factory. Any man with the modern outlook on industrial planning and design would have been called a revolutionary in the middle of the last century. Such men no doubt existed though they were seldom articulate. But in the eighteen-fifties, which began with that magnificent gesture of "Peace and Good Will," the Great Exhibition, and which soon reverted to the more congenial futilities of a war undertaken to preserve the sick man of Europe (who was better dead), there lived and wrote an observant gentleman named Samuel Sidney. He wrote improving works on economics and agriculture, such as *How to Settle and Succeed in Australia*; *Sidney's Emigrant's Journal*; *Railways and Agriculture*, and a stimulating survey of England in its agonized transition from agriculture to industrialism entitled *Rides on Railways*. That, at least, is the title of the book on the title page; its publishers, perhaps remembering Cobbett, called it *Railway Rides* on the cover. It was first issued in August 1851, and although it is ostensibly concerned with scenery, beauty spots (for beauty was already beginning to be segregated in spots) and railway organization, it includes "a glance" at manufacturing towns such as Birmingham, Liverpool and Manchester, and it contains some quite un-Victorian comments on architecture, and an interesting history of industrial design in Manchester.

The power and the glory of railways appealed to Samuel Sidney. He had nothing but praise for the engineering methods that had made in less than twenty years such smooth and efficient organizations possible; but he was not so pleased with the architecture of railway stations. He described Euston, then a young station, and his pungent account reveals the lucid quality of his criticism. After mentioning that the station was an after-thought, as the line was originally to have ended at Camden Town, he said: "The great gateway or propylaeum is very imposing, and rather out of place; but that is not the architect's fault. It cost thirty thousand pounds, and had he been permitted to carry out his original design, no doubt it would have introduced us to some classic fane in character with the lofty Titanic columns; for instance, a temple to Mercury the winged messenger and god of Mammon. But, as is very common in this country,—for familiar examples see the London University, the National Gallery, and the Nelson Column,—the spirit of the proprietors evaporated with the outworks; and the gateway leads to a square courtyard and a building the exterior of which may be described in the language of guide-books when referring to something which cannot be praised, as 'a plain, unpretending, stucco structure,' with a convenient wooden shed in front, barely to save passengers from getting wet in rainy weather." Regarding the internal planning of the station, he held that "comfort has been sacrificed to magnificence. The platform arrangements for departing and arriving trains are good, simple, and comprehensive; but the waiting-rooms, refreshment stand, and *other conveniences* are as ill-contrived as possible; while a vast hall with magnificent roof and scagliola pillars, appears to have swallowed up all the money and all the light of the establishment. The first class waiting-room is dull to a fearful degree, and furnished in the dowdiest style of economy. The second class room is a dark cavern, with nothing better than a borrowed light. The refreshment counters are enclosed in a sort of circular glazed pew, open to all the draughts of a grand, cold, uncomfortable hall, into which few ladies will venture. A refreshment-room should be the anteroom to the waiting-room, and the two should be so arranged with reference to the booking-office and *cloak-rooms*, that strangers find their way without asking a dozen questions from busy porters and musing policeman. Euston station reminds us of an architect's house, where a magnificent portico and hall leads to dungeon-like dining-room and mean drawing-room. Why are our architects so inferior to our engineers?"

Once Mr. Sidney left London his powers of observation were more than respectable. He discovered a queer passion for novelty of design in Birmingham, and found that undertakers were most active patrons of variety in a little-known branch of industrial art, namely, the manufacture of coffin ornaments. "Who is it," he asked, "that examines and compares the ornaments of one coffin with that of another? We never heard of the survivors of a deceased examining an undertaker's patterns. And yet, a house which consumes forty tons of cast iron per annum for coffin handles, stated to the gentleman to whose letters we are indebted for this information, 'Our travellers find it useless to show themselves with their pattern books at an undertaker's, unless they have something tasteful, new, and uncommon. The orders for Ireland are chiefly for gilt furniture for coffins. The Scotch also are fond of gilt, and so are the people in the west of England. But the taste of the English is decidedly for black. The Welsh like a mixture of black and white. Coffin lace is formed of very light stamped metal, and is made of almost as many patterns as the ribbons of Coventry. *All our designs are registered, as there is a constant piracy going on, which it is necessary to check.'* "

When he inspected Manchester he recorded the early and sincere struggles of that manufacturing centre to improve industrial art. In the Mechanics' Institute at Manchester, which was founded in 1824, there was a department of design. "This School of Design," Mr. Sidney wrote, "supported by the Government for the purpose of promoting design as applied to the staple manufactures, and diffusing a general feeling for art amongst the manufacturing community, was formerly accommodated within the walls of the Royal Institution as a tenant, paying a rent, strangely enough, for the use of a building which had ostensibly been erected for promoting art and science! It was not until 1836, that, on the recommendation of a Committee of the House of Commons, active steps were taken to establish in England that class of artistic instruction applied to manufactures which had been cultivated in France ever since the time that the great Colbert was the minister of Louis XIV.

"At Manchester some of the leading men connected with the calico-printing trade and looms of art, established a School of Design within the Royal Institution, where two rooms were lent rent-free; but, as soon as Government apportioned a part of a special grant to the Manchester School, the Committee, who were also as nearly as possible the Council of the Royal Institution, with that appetite for public money which seems incident to men of all nations, all classes, and all politics, voted £100 out of the £250 per annum for rent. This school did nothing of a practical nature, and consequently did not progress in public estimation. The master was a clever artist, but not up, perhaps he would have said not *down*, to his work. A School of Design at Manchester is meant, not to breed artists in high art, but to have art applied to the trades of the city. The master was changed, and, at the request of the local committee, the Council of the School of Design at Somerset House sent down, in 1845, Mr. George Wallis, who had shown his qualifications as an assistant at Somerset House and as master of the Spitalfields school. At that time the Manchester School had been in existence five years, and had done nothing toward its original object. In two years from the time

53

of Mr. Wallis taking the charge, the funds of the school were flourishing; the interest taken in it by the public was great, and nearly half the Institution was occupied by the pupils, while the applications for admission were more numerous than could be accommodated. Under this management the public, who care little for abstract art, were taught the close connection between the instruction of the School of Design and their private pursuits.

"*This* is what is wanted in all our towns. It is not enough to teach boys and girls,—the manufacturers and purchasers need to be taught by the eye, if not by the hand. According to part of Mr. Wallis's plan, an exhibition was held of the drawings executed by the pupils for the annual prizes, which had a great influence in laying the foundation for the efforts made by Manchester at the Great Exhibition of Industry in Hyde Park.

"While matters were proceeding so satisfactorily, the Somerset House authorities (who have since been tried and condemned by a Committee of the House of Commons), proceeded to earn their salaries by giving instructions which could not be carried out without destroying all the good that had been done. The Manchester Committee and Mr. Wallis protested against this *red tapish* interference. It was persisted in; Mr. Wallis[1] resigned, to the great regret of his pupils and manufacturing friends in the managing council.

"The result was that the undertaking dwindled

[1] At the Great Exhibition of Industry of 1851, Mr. G. Wallis, at the suggestion of the Board of Trade, had the management and arrangement of the department of manufactures.

away rapidly to less than its original insignificance,—the students fell off, and a deficit of debt replaced the previously flourishing funds. Out of evil comes good. The case of Manchester enabled Mr. Milner Gibson, M.P. for Manchester, to get his Committee and overhaul the Schools of Design throughout the kingdom.

"Certain changes were effected. The school, no longer able to pay the high rent required by the Royal Institution, was removed to its present site in Brown Street, placed under the management of Mr. Hammersley, who had previously been a successful teacher at Nottingham, and freed from the meddling of incompetent authorities. And now pupils anxiously crowd to receive instruction, and annually display practical evidence of the advantages they are enjoying."

Mr. Sidney was evidently satisfied regarding the health of the School of Design; and perhaps we may attribute its failure to have any permanent influence on the products of Manchester to the fact that manufacturers did not recognize the importance of fresh injections of talent by creative designers, and were content with the mediocre output of docile hacks with a talent for "drawing." Mr. Sidney was one of those rare Englishmen who retained the use of his eyes in a century that had almost lost its sense of sight. He had no illusions about the architectural character of manufacturing towns. Of Leeds he wrote: "The public buildings are not externally imposing, and it is, without exception, one of the most disagreeable-looking towns in England—worse than Manchester; it has the reputation of being very unhealthy to certain constitu-

55

tions from the prevalence of dyeworks." Contrast this description with his view of an industry that began in the eighteenth century. "We may observe, that there is no more pleasant mode of investigating the processes of the woollen manufacture, for those resident in the south of England, than a visit to the beautiful valley of the Stroud, in Gloucestershire, where the finest cloths, and certain shawls and fancy goods, are manufactured in perfection in the midst of the loveliest scenery. White-walled factories, with their resounding water-wheels, stand not unpicturesque among green wooded gorges, by the side of flowing streams, affording comfortable well-paid employment to some thousand working hands of men and women, boys and girls."

Regarding the careless growth of manufacturing towns he was instructive. "Birkenhead," he wrote, "is a great town, which has risen as rapidly as an American city, and with the same fits and starts. Magical prosperity is succeeded by a general insolvency among builders and land speculators; after a few years of fallow another start takes place, and so on—speculation follows speculation. Birkenhead has had about four of these high tides of prosperous speculations, in which *millions* sterling have been gained and lost. At each ebb a certain number of the George Hudsons of the place are swamped, but the town always gains a square, a street, a park, a church, a market-place, a bit of railway or a bit of a dock. The fortunes of the men perish, but the town lives and thrives. Thus piece by piece the raw materials of a large thriving community are provided, and now Birkenhead is as well furnished with means for accommodating a large population as any place in England and has been laid out on so good a plan that it will be one of the healthiest as well as one of the neatest modern towns."

From the account of this contemporary observer we can see the whole confused process of industrialism, its combination of efficiency and muddle, its superbly organized railways, its hopelessly disorganized factory growth, and the incoherent profusion of its productions. The great ability of the English craftsman had provided up to the end of the Georgian period a fine character for the things of everyday use. The comeliness of commonplace objects was destroyed by mechanical production. The educated direction which all design had received from architects was interrupted, and the secondary place to which architects had sunk in the estimation of such clear-sighted critics as Mr. Sidney is suggested by his rather petulant question: "Why are our architects so inferior to our engineers?" The engineer had become the dominant figure of the century, and he was the master technician, the only technician recognized by the manufacturer. The designer had disappeared from society. He had hardly ever appeared in the factory. The sporadic efforts to train designers recorded by Mr. Sidney, and those mentioned in the historical section of the Report of the Gorell Committee on Art and Industry,[1] do not suggest that the

[1] *Report of the Committee Appointed by the Board of Trade under the Chairmanship of Lord Gorell on the Production and Exhibition of Articles of Good Design and Everyday Use*. London, H.M. Stationery Office, 1932. See also the Report on *Design in the Cotton Industry* quoted in Chapter III.

57

designer was regarded as a technician with authority. He was at best a pattern-maker, a malleable draughtsman, the sort of man who could devise on his drawing-board an infinity of variations upon such a theme as Mr. Sidney mentions: coffin ornaments. Machinery could stamp out ornament by the mile. All that was needed to set the machine at work were drawings: and even to-day the confusion of draughtsmanship with design obscures the comprehension of many manufacturers. It is not widely understood that design is more than a trick of the pencil, and that it is produced only by training, thought and artistic judgment. To refer to a studio full of draughtsmen ringing the changes on patterns as the "design department" is a mistake that has been and is being made all over industrial England.

Because of the absence of designers, and because of the facilities for imitation the machine offered, there was a phase of limp adaptation of shapes and patterns originally evolved by hand-craft. These shapes and patterns were misapplied to machine-made things.

Eager to escape such an undesirable compromise, and disliking the individuality inevitably associated with the practice of handicrafts, the present government in Russia appears to be deliberately extirpating hand-craft so that the machine can evolve its own technique of design unhampered by precedents derived from totally dissimilar methods of production. This is a ruthless way of escaping the complicated wastage and debauchery of design typical of English industry during the nineteenth century. Russia, which is being industrialized according to plan, is determined to be logical, and to eliminate any hindrances to the development of machine-craft design. To what extent Russian industrial art will be influenced by contemporary European fashions is not yet apparent. The peasant crafts may disappear. It is believed by those in authority in Russia, rightly or wrongly, that machine-craft is wholly incompatible with hand-craft and that they are incapable of mutual stimulation. But Soviet Russia is not immune from the influence of contemporary styles in design, and public buildings and factories in Russia are already showing the influence of the mechanistic baroque style.[1] No doubt the growth of this style in Russia is strengthened by a genuine belief in functionalism; but the relation of functionalism to a popular style or its confusion with a mode of decoration are old troubles of design. Professor Goodhart-Rendel reminds us that "the ideas now put forward under the ugly name of 'functionalism,' for example, are very old, but it is only lately there has been any widespread attempt to carry them into effect in all their purity. Pugin held them, but the mood of his age flavoured him and them with Gothicism. The mood of our age has flavoured Monsieur le Corbusier and his ideas with engineering, a flavouring that mixes well with the notions it permeates and allows them

[1] "At Moscow there is a 'House of Labour,' deliberately designed by M. Golosov on the model of a dynamo; and the largest and most dominant part of this building is designed as an enormous cog-wheel." Sir Reginald Blomfield, *Modernismus*, chap. iv, p. 53.

frequently to find a more or less acceptable expression in actual building."[1]

Our contemporary obsession with functionalism is at least healthier than the obsession with ornament and with antique styles which complicated every attempt to improve design in the nineteenth century. There were notable precedents for collaboration between manufacturers and designers, and Sir Matthew Digby Wyatt, Slade Professor of Fine Art, in a series of lectures delivered at Cambridge in 1870 discussed the ancient and modern relationship of art and industry.[2] In referring to Wedgwood he said: "It was his delight to work hand in hand with the best artists, and (far in advance of his time) he recognized the commercial value of design as an assistant to Industry. He clearly saw that public interest would be excited by excellence, that cultivation was necessary for the development of taste, and that artists could only properly design for manufacturers, who identified themselves with the operations and specialities of the branches of industry in connection with which they exercised their art.

"Herbert Minton, no less than Wedgwood, spent a long and laborious life in raising the character of the branch of manufacture to which he devoted himself."[3]

The Royal Society of Arts, which was founded in 1754, played an important part during the nineteenth

---

[1] *Vitruvian Nights*, p. 230.

[2] Lecture XIII. These lectures were published in book form under the title of: *Fine Art, a sketch of its History, Theory, Practice and Application to Industry*.

[3] *Opus cit.*, p. 370.

century in the stimulation of design in industry. Prior to the Great Exhibition of 1851, the Society had been active in promoting small-scale exhibitions, and had encouraged experiments in design by awards, which took the form of medals and special prizes. Between 1846 and 1850 recipients of the Society's medals included Minton & Co., and Copeland (pottery); Osler & Co., and Pellatt & Co. (glass); and various other firms whose products included iron castings, carpets, jewellery and safes.[1] In his notes on Birmingham, Mr. Samuel Sidney refers to the glass made by "Messrs. F. & C. Oslers, of Broad Street" and states that the firm has "attained a very high reputation for their cut and ornamental, as well as the ordinary, articles of flint glass. They have been especially successful in producing fine effects from prismatic arrangements. Their gigantic chandaliers of great size, made for Ibrahim Pacha, and the Nepaulese Prince, were the steps by which they achieved the lofty crystal fountain, of an entirely original design, which forms one of the most novel and effective ornaments of the Crystal Palace." He adds: "The manufactory as well as the showroom is open to the inspection of respectable strangers."[2]

The factories of Birmingham and other centres of industry did not lack inventive brains; but much inventive power was devoted to masking functionalism, to "putting on" a little or a lot of "art"; and this

[1] *A History of the Royal Society of Arts*, by Sir Henry Trueman Wood, 1913, p. 389.
[2] *Railway Rides*, p. 80.

idea of clothing a simple piece of machine design with "applied art" prevented the idea of design as we understand it to-day from attaining any sort of effective power. Design in the middle of the nineteenth century was the choosing of the sort of "art" you wanted to "apply"; the selection of embellishment; and "artistic" people were those who knew a lot about the different "styles" of embellishment. Mr. Sidney's remarks upon Mr. Winfield of Birmingham illustrate the results of this approach to design.

"Mr. Winfield is one of the manufacturers in brass whose showrooms are open to the public. He also has claims on our attention for the wise and philanthropic manner in which he has endeavoured to supply the lamentable deficiency of education among the working classes. He holds a very leading position as a manufacturer of balustrades, tables, window-cornices, candelabra, chandeliers, brackets, curtain-bands, and above all of metal bedsteads, which last he has supplied to some of the chief royal and princely families of Europe, besides Spain, Algeria and the United States. In all these works great attention has been paid to design as well as workmanship, as was amply proved both at the local exhibition in 1849, where a large gas bracket, in the Italian style, of brass, with Parisian ornaments, excited much admiration; and in 1851, in Hyde Park, where we especially noted an ormolu cradle and French bedstead in gilt and bronze, amid a number of capital works of his production."[1]

Still dealing with Birmingham, Mr. Sidney records

[1] *Opus cit.*, p. 94.

that "Messrs. Messengers & Sons have one of the finest manufactories in ornamental iron, brass, and bronze, for lamps, chandaliers, and table ornaments. For a long series of years they have spared no expense in obtaining the best models and educating their workmen in drawing and modelling. In their show-rooms will be found many very pleasing statues in gold-colour, in bronze, and copies from antique types of vases, lamps, candelabra, etc."

What did that education of workmen "in drawing and modelling" imply? Did it give them understanding of proportion; did it sharpen their critical judgment of form? Or did it just give them a knack of copying "antique types," and develop their powers of adapting the character of such types to the needs of mechanical production? The illustrated records of the Great Exhibition of 1851 and the catalogues of firms who were manufacturing products that were supposed to be influenced by art, show that by the middle of the nineteenth century the word design as we understand it and as the eighteenth century understood it, had lost its meaning.

Even the work of the Royal Society of Arts, and the awards they made for the encouragement of "art" in industry and the exhibitions they sponsored, could not reinstate design, nor disillusion the manufacturer about the character of the designer. The Society then, as now, was devoted to the promotion of arts, manu-factures and commerce, and its activities were diverse and impressive, ranging from the introduction of medicinal plants to Great Britain to the humanitarian encouragement of inventors to produce mechanical devices which would replace child labour in chimney-sweeping.[1] As early as 1828 an attempt was made to hold a national exhibition of industrial products. George IV. encouraged the exhibition committee with his patronage, and it was proposed to hold annual exhibitions "of new and improved productions of our artisans and manufacturers."[2] The scheme failed to secure adequate support from industry. There were other attempts to organize industrial exhibitions, one in Birmingham in 1836; and in 1844 and 1845, small exhibitions were held in the rooms of the Royal Society of Arts, and although limited in scope and in duration (they only lasted for one evening) they began the movement which led to the 1851 Exhibition.[3]

The Prince Consort was elected President of the Society in 1843, and he took a very broad view of its responsibilities. He insisted that its function was to "improve the condition of the artistic industries of the country" and "had urged on the Society, as its proper work, the encouragement of the application of art to practical purposes."[4] The loss of the idea of what design meant is revealed by that choice of words: *the application of art to practical purposes*. Prizes might be offered for artistic designs; all they produced were applications and variations of ornament, culled from antique sources, by draughtsmen.

[1] Sir Henry Trueman Wood's *History of the Royal Society of Arts* gives a minutely detailed account of the Society's activities from its inception until 1880.

[2] *History of the Royal Society of Arts*, chap. xvii, p. 403.

[3] *Opus cit.*, p. 404.   [4] *Opus cit.*, p. 405.

Yet this growing confusion whicn led to the most repulsive ornamental congestion was accompanied by a splendour of technical achievement in industry. Even Mr. Sidney became almost lyrical about the promise and power of Birmingham and the character of its inhabitants. He wrote: "Birmingham is, in fact, notable for its utility more than its beauty,—for what is done in its workshops, rather than for what is to be seen in its streets and suburbs. Nowhere are there to be found so numerous a body of intelligent, ingenious, well-educated workmen. The changes of fashion and the discoveries of science always find Birmingham prepared to march in the van, and skilfully execute the work needed in iron, in brass, in gold and silver, in all the mixed metals and in glass. When guns are no longer required at the rate of a gun a minute, Birmingham steel pens become famous all over the world. When steel buckles and gilt buttons have had their day, Britannia teapots and brass bedsteads still hold their own. No sooner is electrotype invented, than the principal seat of the manufacture is established at Birmingham. No sooner are the glass duties repealed than the same industrious town becomes renowned for plate glass, cut glass, and stained glass; and, when England demands a Palace to hold the united contributions of 'The Industry of the World,' a Birmingham banker finds the contractor and the credit, and Birmingham manufacturers find the iron, the glass, and the skill needful for the most rapid and gigantic piece of building ever executed in one year."[1]

[1] *Railway Rides*, p. 78.

While technical ability increased and machine design improved, the idea of "applied art" stifled the development of true industrial art. Its chances of healthy evolution were narrowed by one of the consequences of the romantic movement, which took the form of an attempt to revive handicrafts.

"From its inception, the movement for the revival of handicrafts was influenced by romantic antiquarianism. William Morris, the most energetic leader of the movement, was a mediaevalist, a reactionary like Ruskin, and quite unaware of the promise that lurked beneath the smoke-cloud of the factories, of the beauty that machine-craft might bring to the world under proper direction. He only heard the roar of the machines, and saw the dismal mess their masters made of the country, and thought of the lives they stunted; and in their products he saw only a tawdry heaping up of cheap and nasty things, an endless procreation of vulgarity. To Morris and to many hundreds of other artists and designers industry was a cancerous growth, and if it could not be cut out of civilization, then some part of the social body must be kept sweet and healthy by recalling the conditions of the past, and by dreaming of sunny utopias in which no wheel was turned save by the breeze or running water or the power of living muscle."[1]

But even Morris had wistful visions of lost opportunities. In a lecture on "Art and Socialism" he said:

"Those almost miraculous machines, which if orderly

[1] *Men and Buildings*, chap. viii, "Back to the Hand," p. 135.

forethought had dealt with them might even now be speedily extinguishing all irksome and unintelligent labour, leaving us free to raise the standard of skill of hand and energy of mind in our workmen, and to produce afresh that loveliness and order which only the hand of man guided by his own soul can produce; what have they done for us now?"

There was nothing in the so-called "art products" of the mid-nineteenth century that could provide a reassuring answer to a question like that. Only in the work of engineers, concerned with machine design, was there any freedom from borrowed characteristics. Paxton's Crystal Palace which housed the Great Exhibition was the forerunner of the New Architecture that has been advertised so dramatically by its Continental exponents during recent years. It was the beginning of the architecture of steel and glass, and another contribution to this architecture was made in England by Decimus Burton (1800–1881) in the Palm House at Kew Gardens. But even the engineers were hampered at first, and Sir W. M. Flinders Petrie has pointed out in *The Revolutions of Civilization*[1] that "In Mechanics, or the adaptation of long-familiar principles and materials, the full freedom of design was certainly not attained in the earlier railway work. Brunel's tubular bridge, though new, was by no means a perfect adaptation to its requirements. Perhaps Baker's Forth Bridge may be the typical example of freedom from needless restriction, in meeting one of

[1] First published in 1911.

the oldest needs of man with methods and material already well known, apart from fresh discovery."

But in all this machine design which was made possible by new metallurgical knowledge Morris could find nothing for the people, nothing to replace the lost common art of England. Had he given his great abilities to solving the problem of relating design to industry a real and vigorous industrial art might have arisen in the last quarter of the nineteenth century. But instead:

"He turned to the past with a sigh of relief: there he found refuge from the clamorous realities of his own century. He looked back, not to the great civilization of the Renaissance: that was too intellectually lordly at the expense of the men with tools in their hands and creative emotion in their hearts: so back to the days of the mediaeval guilds he went in spirit. He looked back, and by the exercise of that convenient editorial faculty that is essential to the maintenance of a perfect state of self-deception he ran a mental blue pencil through all the miseries and limitations, the ills and frantic fears of the Middle Ages. The age of the mediaeval craftsmen became for him an illumined manuscript, unsoiled by ugly facts, shining with bright colours and packed with inspiring texts. He early abandoned the idea of becoming an architect, partly because the architect of the mid-nineteenth century was a man of drains and drawing-boards who dealt in 'styles,' but chiefly because the arts and crafts had dissolved partnership with building. He wanted to see houses and cities growing into beauty under the hands of craftsmen: he wanted wood and stone to be carved freely and surfaces to carry a burning splendour of decoration.

"The house that Philip Webb built for him at Upton, in Kent, was a small-scale model of this great ambition. It was called the Red House, and it gave a new shape to the romantic movement in domestic architecture. It was a two-storied house with walls of red brick and a high-pitched roof of red tiles. The plan was L-shaped. (For some reason or reasons unknown, it was planned so that the sitting-rooms, the dining-room, the drawing-room, and the hall, all faced north.) There was a careless and comfortable independence in its character: its oriel windows and gables were unostentatiously romantic; and within, the furniture and decoration were strongly individual. It possessed the 'quaint richness' that Ruskin had applauded; but its antiquarian flavour was incidental, for it was a sincere attempt on the part of some singularly gifted people to solve an architectural problem from a particular point of view.

"Morris and the group of artists who shared his sympathies felt that architecture should arise naturally and joyfully from a revival of the crafts, and that the work of a brotherhood of craftsmen must transcend the tyrannical harmonies imposed by the Renaissance. Hitherto the Gothic revival had been a thirsty search for picturesque forms, which, in spite of all Ruskin's eloquent directions, was usually satisfied with copyism. Morris tried to resurrect the creative spirit of the men who had made the mediaeval abbeys and guild-halls, and he devoted his abounding energy to mastering a number of crafts, not as an artistic dabbler, but as a skilled craftsman.

"His personal powers were bewildering in their variety and perfection. He was a poet of a high order—a teller of tales, whose prose unrolled like some glowing tapestry, with the story vividly depicted in rich colours; and he was a great decorative artist.

"After the building of the Red House, Morris and his friends realized that every branch of decorative and applied art in England was in a state of advanced decay. It was impossible to buy well-designed furniture, fabrics, or wallpapers, and it was to elevate standards of domestic design that the firm of Morris and Company was founded in 1861. Philip Webb, Burne-Jones, Rossetti, Ford Madox Brown, Faulkner and Marshall were associated with Morris in this venture. This firm was prepared to undertake church decoration, carving, metalwork, stained glass, and also to deal with wallpaper, chintzes, carpets, and furniture.

"Unfortunately the practical expression of the 'back to the hand' doctrine was expensive. Handicrafts could be revived, or their extinction delayed, but the craftsmen could not live upon the joy of work alone, and the cost of living was higher in the nineteenth century than it was in the Middle Ages. Consequently Morris perforce found himself working for that relatively tiny section of the community that had both riches and artistic perception. The lives and homes of the common people were untouched, and common

art—that dear popular possession of mediaeval England—was still unrestored to the multitude.

"Presently the costly products of organized handicraft were imitated by industry. Morris had, quite
unintentionally, started a vogue for 'hand-made'
articles, and the manufacturers had a new label for
their wares. The machine was equal to the demand
for 'art' products; and the handicraft note was
admirably simulated by speckling metalwork with
mock hammer marks, leaving woodwork rough and
heavy; and, where no external evidence of hand-
work could be faked, emphasis was laid by shopkeepers
on the 'quaintness' of the form or the ornamentation
of the articles they had for sale.

"A few crafts had been precariously preserved by
the influence of Morris; but in providing opportunities
for craftsmen to work, he had omitted to furnish them
with the right customers. It was galling to have one's
activities supported by the 'arty' rich, while the
'people' (as Morris thought) were starving for colour
and gaiety and carving and folk-songs amid the reek
of the factory chimneys and the clangour of machinery.
Before comon art could come back to their lives a
social revolution would have to take place; so Morris,
without bothering his head about any economic
quibbles, became a socialist. Meanwhile his teaching
gave a fresh impetus to the romantic antiquarian
movement, and established an exaggerated reverence
for hand-work and handicraftsmen. Therefrom arose
two evils: firstly, blind admiration of the antique,
which begot the curse of sterile imitativeness and

atrophied all critical judgment of design; and secondly, the intolerance of the creative artist for machine-craft which has robbed modern industry of immeasurable advantages, and has made the designer a stranger to businesses where he should most properly be a partner."[1]

The Arts and Crafts Movement started by William Morris impeded the proper development of industrial art in England, because it captured the interest of so many artists who might otherwise have been attracted to problems of industrial design, and because its preoccupation with the past increased the readiness of manufacturers to imitate traditional forms.

Sir W. M. Flinders Petrie has given 1890 as the date for the close of archaicism in mechanics.[2] Although as

FIG. 12—*The Forth Bridge: an example of the new machine architecture*

[1] *Men and Buildings*, chap. viii, p. 136–140.
[2] *Revolutions of Civilization*, chap. v, p. 97. (Third Edition.)

mentioned earlier in this chapter he suggested that the Forth Bridge might represent a boldly free use of materials and mechanical methods, he qualified the observation by suggesting that further work might show that archaicism had even clung to the Forth Bridge.

But there were very few plain statements of design in the nineteenth century, because, as Mr. and Mrs. Clough Williams-Ellis have pointed out in *The Pleasures of Architecture*: "we are afraid of size, afraid of plainness. We love sweetness and mistrust strength." And they ask, "Who but the English would have 'dolled up' that magnificent engineering feat the Tower Bridge in the lace flounces of prettified Scotch baronial architecture, utterly destroying its dignity, and making it a silly and coquettish chatterbox?"[1]

But the "dolling up" of the Tower Bridge was just another example of the idea that art was "applied." The transporter bridge at Runcorn is a plain statement of machine design. So is the transporter bridge at Newport, Monmouth; so are the towers that carry the power lines of the Grid. If the Central Electricity Board had done its work in the 'seventies or 'eighties, and money had been abundant, it is conceivable that as an alternative to burying the cables underground, there might have been a scheme for making the steel lattice towers that carried them overhead, look like something quite different. "Art" might have been "applied" to make them resemble gargantuan rustic arbours, for

[1] Chap. x.

FIG. 13—*The Tower Bridge: an example of the new machine architecture in fancy dress*

example, with imitation roses (in enamel) crawling over them. Such was the taste of the nineteenth century. It is a form of taste that still persists.

His Grace, the Archbishop of Canterbury, at the Royal Academy banquet in 1934 referred to the Battersea Power Station. He said that it was the one industrial building which they might have supposed would have withstood the advance of art. "Yet the genius of Sir Giles Scott had invested even it with a real nobility of art." Referring to the projected exhibition of industrial art at the Royal Academy he expressed a hope that if it could "inspire the masters and men of the industries to take a new interest in their workmanship, and believe that even beauty could have a marketable value, they would have some consolation for living in an industrial age."[1]

The attitude of mind disclosed by these views is that of the escapist who falls back on disguise to alleviate his distaste for the commercial machine age. Earlier in his speech the Archbishop referred to one alliance between art and industry and he said "they had seen the services of great artists enlisted, and the railway stations gradually converted into excellent picture galleries." Again the idea of disguise protrudes from this statement. The covering of ill-designed and untidy railway stations with posters by Royal Academicians is like giving a false and transitory appearance of health by means of artificial sunburn or bronze powder to a man who is suffering from a grave organic disease for which skilful surgery is the only cure. The application of "art" can never remedy an absence of plan.

By the end of the nineteenth century the form of nearly all industrial productions was obscured by ornament, and *such* ornament. It was sometimes quite difficult to recognize the original function of an object after the hack pattern-makers had done their tricks. But it was an age of ornament, and it was consistent in its copiousness. Sometimes an old film of the first decade of the present century will illustrate forcibly something which few people remember, and that is that thirty years ago, clothes, buildings, vehicles, rooms and people were all of a piece. The background of the street scenes is composed of sore-looking Gothic

[1] Report in *The Times*, Friday, May 4, 1934, p. 9.

and bulbous classic façades, all overlaid with a thick coating of sugary decoration. And the bulbous and expanding draperies of the women, and the unwieldy roofing materials they used to protect a "hayrick head of hair," matched this background perfectly. We have inherited the background and forgotten the clothes. We have inherited the unfortunate tendency to overlay everything with ornament so that the purpose of an object is often completely obscured. Really beautiful processes of machine-craft are warped and spoiled because the manufacturer and the retail buyer who controls the distribution of the manufacturer's goods are uneducated in design, uninfluenced by any contemporary movements, and still sharing with many people who have some claim to culture the view that "art" is something extraneous, something to be "applied."

The Arts and Crafts Movement under the influence of Morris did re-establish a respect for functional fitness; but it was entirely concerned with rehabilitating hand-crafts, and its influence upon industrial production was negligible. It attracted a large number of good designers, it produced a generation of artist-craftsmen, including such men as Ernest Gimson and Sydney Barnsley. It produced a number of misconceptions about art that were in every way as damaging to lucid thinking about the subject as the idea of "applied art." The confusion of competent craftsmanship with ability to design was occasionally set forth in such striking and musical phrases as those employed by the late Professor W. R. Lethaby, like "Art is thoughtful workmanship." But although Professor Lethaby exalted competent workmanship, he made many wise and clear statements about the function of machine-craft. Most misleading is his statement that "art may be thought of as *the well-doing of what needs doing.*" Most illuminating is his statement that: "Although a machine-made thing can never be a work of art in the proper sense, there is no reason why it should not be good in a secondary order—shapely, smooth, strong, well fitting, useful; in fact, like a machine itself. Machine-work should show quite frankly that it is the child of the machine; it is the pretence and subterfuge of most machine-made things which make them disgusting."[1]

But criticism of the Lethaby type was conceived in what may be called the handicraft revivalist spirit.

The effect of the rebellion of the artist-craftsmen and their supporters against the general unfitness of design in the nineteenth century, produced some interesting results and some beautiful things, particularly in furniture. We may consider the work of Ernest Gimson, the greatest artist-craftsman produced by the Morris Arts and Crafts Movement, who was responsible for the making of many pieces of beautiful furniture:

"Gimson was a craftsman endowed with the ability of a designer. He was not just a designer who dabbled in handicraft and knew how to employ other craftsmen. It is important to recognize his ability as a designer, for it is sometimes supposed that an accomplished craftsman is by virtue of his manual dexterity

[1] *Form in Civilization*, by W. R. Lethaby, chap. xix, p. 211.

a designer. The craftsman left to his own common sense, may devise something that is fit for its purpose, but he may over-decorate it like any savage; he may be unaware of innumerable opportunities for refining the proportion of various members; he may achieve a solid straightforwardness, a rustic simplicity, but in the absence of a continuous tradition of furniture making to nourish his invention and provide him with guiding precedents, he must improvise, and, unless he has the selective and inventive skill of a designer, his improvisations may be discords. By mastering the craft of woodworking, Gimson, the sensitive and accomplished designer, brought to furniture making the individual genius it had lacked since the death of Sheraton. After Sheraton there had been no great names associated with English furniture. The supply of men with ideas disciplined by a craftsman's training had dried up: there were plenty of drawing-board men, and the Victorian age is grim with the indiscretions of their taste. Sheraton was a craftsman before he started publishing books on design; Hepplewhite was a craftsman; so was Chippendale. Gimson's affinities with the traditional English craftsman are indisputable; and posterity will probably single out his name when it seeks for evidence of early twentieth-century ability in furniture design. But he did not follow the line of fashionable designers that ended with Sheraton. His work continued, unconsciously, the developments that had been suspended by the restoration of Charles II.

"He was only concerned with hand-craft methods.

He took no assistance from the machine age. He felt that any compromise with mechanical production was impossible. 'Let machinery be honest,' he said, 'and make its own machine-buildings, and its own machine-furniture; let it make its chairs and tables of stamped aluminium if it likes: why not?' So aiming at a mastery of the traditional methods of woodworking he gave to a largely uncaring world beautiful examples of original furniture. He made chairs with turned legs and rails and ladder-backs, and spindle-backs and rush-seats, displaying his ability for apt decoration; also cabinets and chests and sideboards in walnut and oak, with veneering of burr-elm, with inlays of holly and ebony, cherry, ivory, bone and mother-of-pearl. His chairs were of oak, ash, yew, walnut and elm. All the gifts of English wood, those riches of colour and marking, he used, reviving forgotten knowledge of the decorative quality of such materials as yew and elm; employing them with an ever-widening comprehension of their flexions, of their willingness to be coaxed into comely shapes when they were wisely chosen, part by part, for the work they had to do as components of a chair or table.

"There were no classic mouldings on Gimson's furniture; no trace of an architectural heritage from the Georgian age. He eased angles with chamfers. Cupboard doors were gently raised with fielded panels, He created individual pieces of furniture. The suite, which was originally a gracious invention, spoiled by the dull wits of the Victorian furniture trade, did not inspire him.

79

"Gimson by studying crafts that had survived a precarious existence in the nineteenth century, was able to recapture the natural aptitude of the English woodworker for appropriate ornament. Craftsmen with a tradition behind them could be trusted to use decoration wisely; such wisdom, it may be repeated, does not come from executive skill alone. To-day the only woodwork that maintains a long tradition of ornamental treatment is that used in costers' barrows and farm carts. The painting of such barrows and carts, the shaping of their structural members, with the wood nicked and rounded and the angles chamfered, illustrate pre-Georgian decorative survivals.

"Sydney Barnsley was a designer and maker of furniture of the same school as Gimson. It is difficult to avoid the use of terms such as 'school' in describing the work of a number of individual designers who were solving problems and practising crafts in different parts of England during the first two decades of this century. What they achieved was virtually a new start for hand-craft in this country. The lesser men in this movement produced simple crudities which suggested the bleak forms of the American 'Mission' furniture, bare and solid wooden shapes, unsoftened by any respect for visual comfort, primitive and profoundly unimaginative. From this sort of furniture arose the 'cottage' style, which ran its course in pre-war days and continued after the war with certain concessions made to a desire for 'colourful' gaiety. Gimson richly developed the best that was in Morris's teaching, and ultimately his work slightly stirred the ideas of the furniture trade, for, some years after his death, several manufacturers experimented with simple oak furniture that was more sophisticated than the cottage style and which bore traceable resemblances to some of Gimson's designs."[1]

Although various branches of hand-craft might be temporarily rejuvenated as a result of the Arts and Crafts Movement, industrial design was not seriously affected. But the subject continued to command the attention of experts and critics. In 1914 a joint scheme was framed and sponsored by the Board of Trade and the Board of Education for establishing an organization which was to be called the British Institute of Industrial Art, and which should be devoted to promoting industrial art. "The scheme which was framed in 1914 was eventually launched in 1920, and the Institute received a Treasury grant in its initial stages. Unfortunately the economic crisis of 1921–1922 put an end to any further Government support and correspondingly curtailed the scope of the Institute's programme and achievement. Thrown on its own resources and supported solely by voluntary subscriptions, the Institute has nevertheless organized Industrial Art Exhibitions in London, the provinces and overseas, and has also carried out valuable research work, besides maintaining within the building of the Victoria and Albert Museum a small permanent collection, on

[1] *English Furniture* (Black's *Library of English Art*), by John Gloag, chap. vii, "Furniture Design Under the Antique Dealers and Artist-Craftsmen," 1900–1920, pp. 147–150.

the lines originally contemplated, but on a more restricted scale."[1]

In 1915 the Design and Industries Association was formed. The reason for its existence was concisely stated by Mr. C. H. Collins Baker in the introduction to the first Year Book published by the Association in 1922. He wrote: "Without being too historical we will state that the D.I.A.—the Design and Industries Association—was founded in 1915 by a handful of practical enthusiasts to combat the unpractical influences in British design and industry. They discovered that British things—furniture, textiles, pottery, printing, and so on—were often poor because they were not designed and constructed principally to do their job with maximum efficiency. Hence arose the chief article of their creed—'Fitness for purpose'—and the courage to restate that, if a thing were unaffectedly made to fulfil its purpose thoroughly, then it would be good art. Thus, at one blow, the formidable superstition that real art depended on the elaboration and disguise of multiplied ornament was challenged."[2]

Mr. Collins Baker pointed out that the enthusiastic founders of the D.I.A. did not shirk the facts of contemporary life. "They did not," he said, "indulge the view that machinery, a vulgar, vile affair, caused all our modern ills. If they had a nostalgic hankering for the stage coach and the manuscript, they faced the fact like men, that steam and electricity, the printing press and the typewriter would endure and were

[1] Gorell Report on Art and Industry, 1932.
[2] *Design in Modern Industry*, the Year Book of the D.I.A., 1922.

82

capable of true service. They had the larger vision which perceives that the disease of modern design and industry was due not to machinery but to the imperfect comprehension of its limitations and possibilities. They saw that if modern design were frankly conditioned by the special capability of the *de facto* agent of production, fine art and craftsmanship were compatible with machine-made goods."[1]

The Design and Industries Association with its slogan "Fitness for purpose" did a considerable amount of educational work in the staging of exhibitions, in the arranging of lectures and in organizing travelling exhibitions of illustrations and models which showed good examples of industrial design. In its early years it was apt to confuse industrial design with the practice of the artist-craftsman, and in its literature it displayed a tendency to illustrate the functional fitness attained by the products of hand-craft instead of concentrating wholly upon the improving of industrial design. But the influence of the Association was far-reaching, and it numbered among its members such men as Frank Pick, the present Chairman of the Board of Trade Council on Art and Industry,[2] and one of the greatest patrons of machine design and industrial design that this century has known. As Joint Managing Director of the London Underground Railway Group, and as Vice-Chairman of the London Passenger Transport

[1] *Design in Modern Industry.*
[2] The formation of this Council was announced by Dr. Leslie Burgin, Parliamentary Secretary to the Board of Trade, at the Annual Dinner of the Design and Industries Association, December 4, 1933.

Board, he has in many visible directions been able to influence the character of London. The Underground Railway system, a century and a half ahead of its time, is an example of what railways might be like, if their rolling stock, their equipment, accessories, and station architecture were in the hands of competent designers. The late Sir Lawrence Weaver, for some years the President of the D.I.A., was another influential patron who helped to apply to industrial production the ideas propagated by the D.I.A., and he was personally responsible for introducing many manufacturers to designers, and for founding a number of partnerships of this kind which resulted in the betterment of industrial design.

Far more attention was given to the subject of industrial design after the war. The influence of the British Institute of Industrial Art and the D.I.A. was apparent at the British Empire Exhibition, and when the standard of display and the character of such things as notices and direction signs at that exhibition are compared with those at the pre-war White City Exhibitions, a relatively enormous advance is observable.

The Royal Society of Arts was not inactive in this post-war period, and in 1923 they started an Annual Competition for Industrial Designs, and over a period of ten years the sum of £5,000 was expended on this movement.[1]

[1] In the report on the competition of industrial designs which was issued in 1933, it was stated that "the total number of competitors who entered for the various sections of the Competition

In announcing that the competition was to be discontinued, the Council of the R.S.A. stated that they believed the competition had had a powerful effect "in directing into more practical channels the work of the schools of art." In the report on the competition it was stated : "In far too many cases there is no vital connection between Schools of Art and Industry. Design is taught 'in the air' as it were, and without reference to its practical application. Nor, in most Schools, is any attempt made to find employment for their students when their course is finished. Over and over again candidates in the Competition have come to the Society asking for advice as to how to obtain work. They are pathetically ignorant of the ways of the world and have not the remotest idea either as to what kind of firms are most likely to be useful to them or as to how they should proceed to get in touch with them. Something has been done to guide these lost sheep into the way of employment, both by the Competition itself, and the exhibition of selected works, by which designers of merit are brought to the notice of manufacturers, and also by the Employment Bureau, where are kept was 1,131. Of these 724 were students of Schools of Art, and 407 non-students.

"The number of designs submitted was 2,623, divided as follows :

Architectural Decoration	306
Textiles	1,286
Furniture	88
Book Production	118
Advertising and Commercial Art	760
Miscellaneous	65
Total for all sections	2,623"

85

the names of designers in search of work. By these means a good many permanent posts and a large number of commissions have been found for successful candidates."

Although sporadic and disinterested efforts have been made by independent bodies and private individuals to improve standards of design in industry, those in control of industrial production have not until recently shown any disposition to understand that the employment of good designers and the improvement of design as a consequence is a good business policy.

In November 1933 at the Royal Society of Arts dinner at which the Royal Academy Exhibition of Art in Industry of January 1935 was formally announced, His Royal Highness the Prince of Wales said: "Our industries have been developed by scientists and technical experts of all kinds, they have been identified with every branch of industry where they would make for improvement in production or distribution. All forms of experts, except artists, have been employed because manufacturers have not recognized how the artist can sometimes help in the design and the consequent sale of a commodity."

That neglect has determined the character of British factory-made goods from the beginning of the commercial machine age to the present day. Engineers have competently solved problems of machine design. Result: British railway engines, the motor-buses and trolley-buses of the London Passenger Transport Board, and, in industrial architecture, the Forth Bridge, the fans of the exhaust shafts in the new Mersey Tunnel and the towers that carry the power lines of the Grid. But the designer has been the missing technician, and the manufacturer has solved his problems of design with the aid of hack draughtsmen. Result: Flashy imitation of antique things in metal and wood. Old patterns repeated *ad nauseum* in textiles and wall-papers. Deliberately cranky shapes in utensils, twisted and riven to give a touch of "novelty" and to live up to that lamentable label "artistic." The commercial machine age has until recently dispensed with design. It will not be economically possible to continue such neglect.

Occasionally some large retail establishment is tempted to flirt with the problem of design in industry; but it is usually the occasion of an exhibition of the work of some designer or group of designers. It is a transitory stunt; encouraging, but as a contribution to the development of industrial design, impermanent.[1] The Exhibition of British Industrial Art in relation to the home which was held in London at Dorland Hall

[1] The experiment made by Messrs. E. Brain & Co., Ltd., of Stoke-on-Trent, is a brilliant exception. Early in 1933 this firm of potters decided to produce some real contemporary work, and they invited the following artists to carry out designs for China tea-services and earthenware dinner-services: Frank Brangwyn, Laura Knight, Ernest Proctor, Mrs. Dod Proctor, Duncan Grant, Vanessa Bell, Paul Nash, John Armstrong, Ben Nicholson, Barbara Hepworth, Allan Walton, Albert Rutherston, Graham Sutherland, John Everett, Milner Gray, Moira Forsyth and Gordon Forsyth. The results of this experiment were exhibited by a great London retail firm in the autumn of 1934. See Plates XIV, XV and XVI.

during the summer of 1933 was a stimulating illustration of what a number of able designers could do with modern materials and processes; but its exhibits were largely designed for this special occasion; they were not representative of contemporary industrial art; and in consequence there was an air of prophetic irreality about that exhibition, as indeed there must be about all exhibitions of industrial art that are not exclusively devoted to showing the results of design in modern factory-made goods. Nevertheless, exhibitions such as the 1935 Royal Academy Exhibition of Industrial Art, and its small-scale forerunner at Dorland Hall in 1933, perform a valuable function in stimulating the interest of the public in design, and in bringing manufacturers into contact with some of the best minds in the modern movement.

CHAPTER III

THE PRESENT STATE OF DESIGN
IN INDUSTRY AND
THE ECONOMIC PROBLEM OF PATRONAGE

INDUSTRIAL art to-day has some chance of developing a distinctive character of its own. The modern movement in design, although still largely misunderstood and vigorously abused by ladies and gentlemen with a late-Victorian outlook, does encourage designers to make use of the mechanical facilities of the age they live in. The modern designer is not content with the barren achievement of functional fitness. It was necessary for the modern movement to pass through a phase of concentrating upon fitness for purpose, because as we have seen, utility was generally ignored or outraged for the sake of ornament in Victorian and pre-war times. There followed a period when abstention from the impulse to ornament an object was considered a virtue in a designer; and there arose from this period of disinfection a wholly fallacious idea that fitness for purpose automatically achieved beauty. As Mr. Roger Fry pointed out in his Memorandum which was an appendix to the Gorell Report on Art and Industry: "Though such a theory would at least spare us the horrors of futile decoration it would none the less prevent any real development of artistic design. It is true that the best designs often take the functional purpose of an object as a point of

89

departure, but the aesthetic satisfaction given by a beautiful design is quite distinct from the pleasure of recognizing functional adaptation. Good architecture must always remain distinct from good engineering and this principle holds equally in the design of objects of daily use."[1]

To designers of limited fertility, the slogan "Fitness for purpose" provided an opportunity for practising a crude form of functionalism, unrelieved by any concessions to ancient human needs, of which the desire for ornament is perhaps as old as civilization. Various other slogans connected with design, and particularly with architectural design, were circulated to comfort bright young people who wanted to be aesthetically modish without the fatigue of thinking. Critics and designers would murmur "out of the ground into the light," or "the house is a machine for living in," and the works of Monsieur le Corbusier encouraged an exaggerated respect for industrial architecture, so that otherwise quite intelligent people were prepared to lavish extravagant admiration upon such squalid utilitarian objects as gas-holders and factory chimneys.

Architecture is the human activity which above all others takes the impress of its time and preserves the character of each period for the enlightenment of posterity. Architecture will not only record the social and industrial revolutions that have taken place since

[1] *Report of the Committee appointed by the Board of Trade under the Chairmanship of Lord Gorell on the Production and Exhibition of Articles of Good Design and Everyday Use.* H.M. Stationery Office, 1932, p. 44.

1900; it will also record the revolution that has taken place in building and in structural technique, in the equipment of buildings and in the larger organization of architecture which comes under the heading of town planning and street equipment. Architectural design is beginning to influence industrial design, particularly in such things as electric fittings for light and heat.

We are to-day acutely aware of the importance of planning in our buildings, our towns and our cities. It is that awareness, that livelier apprehension of the importance of common sense in everything connected with equipment, that makes the world we live in so different from the Victorian and the pre-war world. This understanding is a new force in design, a new and potent influence in life. In 1907 Professor Sir W. M. Flinders Petrie wrote a disturbing book called *Janus in Modern Life*. It showed the maddening effect inefficiency in planning then had upon every acute and critical mind. The author denounced "the strange lack of thought and adaptability in common matters of everyday life." He listed a number of complaints to support this statement, and in view of later achievements which we now regard as commonplace, his list of defects in everyday life has exceptional interest. He said: "The daily loss of time, and cost in trivial matters, which affects thousands of persons, makes a heavy tax on the whole. For instance, such a simple matter as putting the offices of a terminal station at the ends of the platforms is still ignored at many termini; the name of a station is often hard to find, and is never once put up in most termini; the price of a ticket is often not to be discovered; the right types of carriages are only now being tried; after persevering in a wrong form for two generations." (Mr. Samuel Sidney's inventory of Euston's defects is recalled by this.) "In the streets the same lack of sense is seen in the immense omnibus system, which is difficult to use, especially for strangers, owing to the lack of numbered routes and conveyances. It has been officially decided that the numbering of routes and omnibus is beyond the powers of the London County Council; and we must be compensated by the pleasing reflection that something at least is too hard for that body. The thoughtless edict however was enforced that every vehicle must carry a white light in front, and all the distinctive colours of the tram-car lights were abolished, causing great inconvenience at night." (The achievements of the old London General Omnibus Company, now part of the London Passenger Transport Board, became even more impressive after this description of Edwardian chaos.) "Even in the most recent appliances the same dulness is shown; electric fans are commonly placed where they only stir foul air, and not where they draw in fresh air or expel used air. The whole lighting system still throws away two-thirds of all its cost by lighting sky and walls as much as streets. In every direction it seems hard to believe that five minutes' thought has been given to matters costing thousands of pounds. If we trace such a mixture of design and of chance in any other subject it would lead to some curious speculations on the implied limitations of the directing Intellect. And in private matters it is the same; the extraordinary blunders and oversights in common trade work show that the most obvious details have not had a minute's real thought given to their arrangement. The result is an accumulation of difficulty and muddle which cripples, if not destroys, the purpose of the work. This persistent dulness, and incapacity for management and design, shows a defect of character which is a heavy detriment to the whole community."[1]

We have passed through a great period of make-believe in architecture to a time when we are making tentative acknowledgements of the possibilities the structural revolution has thrown open to us. The revolution which changed the structural character of buildings had taken place before the twentieth century began. Important mechanical services for buildings, lifts, electric light, highly efficient ventilating systems, and the perfection of heating systems such as the panel system, all helped to make modern architecture independent in character; and, before the war, a few voices, crying, of course, in the wilderness, acknowledged this disrupting fact. In 1910 Professor Lethaby, in a paper delivered before the Royal Institute of British Architects, said: "The method of design to a modern mind can only be understood in the scientific, or in the engineer's sense, as a definite analysis of possibilities—not as a vague poetic dealing with poetic matters, with derivative ideas of what looks domestic, or looks farm-like, or looks ecclesiastical—the dealing

[1] Chap. ii, pp. 16–17.

with a multitude of flavours—that is what architects have been doing in the last hundred years. They have been trying to deal with a set of flavours—things that looked like things but were not the things themselves." He concluded his paper by saying, in that age when classic façades and "free" interpretations of Jacobean and semi-baronial elevations were being unpinned from hundreds of drawing boards and tacked on to hundreds of steel frames in streets that had a tradition of dignity, that "The living stem of building-design can only be found by following the scientific method."

The war shook things up so thoroughly that immediately afterwards, under the powerful and quite natural desire to get back to pre-war conditions, we find a pseudo-Georgian revival, planting here, there and everywhere in the country agreeable buildings, and we find the banks fresh from their ponderous amalgamations, expressing their dignity and solidity in classical terms. The five orders of architecture have become for the big five, the insignia of impeccable respectability; rather like old school ties. Meanwhile, on the Continent things were stirring. The first outward and visible signs of Continental influence came not from France, but from Sweden, for the 1923 Exhibition at Gothenberg had a profound effect upon English taste. Sweden was discovered by England; and we can never be quite sure whether the credit for the discovery belongs to that crusading critic, Mr. Clough Williams-Ellis, or to the late Sir Lawrence Weaver. Unquestionably the public exposure of our own forms of taste, industrial, traditional and artistic, which took place the following year at Wembley was not uninfluenced by the small-scale Swedish example of Exhibition *décor* which had preceded it. The next notable example of foreign influence operating upon English taste in architectural and industrial design was the 1925 Paris Exhibition. The first observable effect of this event was upon the furniture trade, which has not yet recovered from this injection of decorative inspiration. But these particular foreign influences have been superficial although their effects are still with us. The real disturbance began with the importation of sermons by functional Puritans (or alternatively puritanical functionalists) like Monsieur le Corbusier. Now, owing to the effect of these sermons, and the examples of building they have produced abroad, we are in that uncertain and difficult position which attends the possession of "an open mind." Into this open mind all the modern materials, some being of respectable pre-war standing, have been dropped. The results are confusing, especially as architecture and industrial design have been "taken up" by so many semi-intellectuals that the intelligent use, the scientific and logical use of the building materials that are to-day available, is often tainted with a desire to shock the eye by some bleak drama of form that is inexcusably uncivilized. But we are shaking off make-believe. It would have been impossible for any architect to have produced a building like the new *Daily Express* building, or Mr. Joseph Emberton's Universal House before the war; not because the materials for such forms of architectural design were not all available,

95

but because it would have been completely impossible to persuade any patron to accept such a design. Also, that free and vigorous approach to the logical use of materials had not developed in pre-war times. To-day we have got efficient and economical systems of heating, of ventilation, of lighting and of power in buildings. We have got all manner of refinements of materials such as plywood which since the war has attained many new forms; and glass which has multiplied its functions, and acquired some exceptional properties, such as a quite untraditional toughness and the power of admitting natural ultra-violet radiation. In the metals, stainless steel and various plated finishes have made it possible to change the whole point of view about the use of bright expanses of metal. No longer have such displays of brightness to be related carefully to the labour available for keeping them in condition. Cellulose paint, the technique of spraying paint, and innumerable plastic, metallic and fire-hardened materials, and an enormous array of patent composition boards, have made it perfectly possible for the modern building to be a light and airy skeleton of metal, weighing less, costing less, and looking far more intelligent and being in design and arrangement far more practical, than the traditional type of steel building veneered with ponderous stage scenery of expensive stone. In architecture we have arrived at this healthy and promising acknowledgement of the activities and technique of our own century. And fortunately the designers who are working out the character of the modern movement are not attempting to adapt traditional ideas to the new mechanical processes and structural methods of this commercial machine age. Their approach to every problem of design represents not a new way of thinking (for common sense is not a new invention), but a lucid way of thinking. They use the new materials frankly and openly in architecture, and it is inconceivable that any architect whose greatness is acclaimed by posterity and proven by his works would have done otherwise. Is it likely that Sir Christopher Wren or Sir William Chambers would have been content to ignore the new rhythms made possible by the structural revolution, fond as Wren was of a "good Roman manner" of building, and meticulous as Sir William was regarding the formal details of classic architecture?

The real character of the modern movement in design, and the discrimination of its most intelligent and imaginative exponents are unobserved by those critics who still think that modernism is merely abstinence from decoration. This sort of remark is typical of the reactionary school of criticism: "One does not want to live either in a conservatory, or in rooms which appear to be suggested by the operating rooms of a hospital." Thus Sir Reginald Blomfield in his most stimulating attack upon the modern movement in design, which he called *Modernismus*.[1] Again, Mr. John de la Valette, in the introduction to *The Studio Year Book of Decorative Art*, 1934, refers to "hospital-cum-factory furniture," and is most encouraged by his belief that in 1933 "women have begun to reassert

[1] *Modernismus*, chap. iv, "The New Architecture," p. 57.

themselves, and to go against the dictates of the solemn kill-joys, the aesthetes, the theoretical designers and, worse than all the lot, the many who write about art." Mr. de la Valette expresses the view that "the reaction against superfluous objects and aimless ornament has gone the full length, until anything falling short of the hygienic standards of a hospital dormitory is looked upon with suspicion."

Even a writer of the intellectual standing of Mr. Aldous Huxley betrays the same misconceptions about the modern movement in much the same phrases. In one of those richly discursive paragraphs in *Beyond the Mexique Bay*, he says: "For us, to-day, the highest luxury is a perfect asepsis. The new casino at Monte Carlo Beach could be transformed at a moment's notice into a hospital. . . . The Wagon-lit Company's latest coaches are simply very expensive steel nursing-homes on wheels."

But the period of asepsis is past; the competent designers have left it behind them. Of course there is a time-lag in design. We have mentioned that the furniture trade is still under the spell of the *Exposition des Arts Decoratifs* of 1925. But one cannot help being a little surprised at Mr. Huxley for being so out of date with his information.

But while such critics as Sir Reginald Blomfield, Mr. John de la Valette and Mr. Aldous Huxley are expresssing their various reactions to the manifestations of the modern movement, the people in control of industry are largely unaware that there is such a movement. To them the word modern means "jazz" or "futuristic" stuff. Because of this, and because adequate payment for good designs would be regarded as a fantastic extravagance by many manufacturers, the talents of many designers who are taking part in the modern movement are withheld from industry. At no time in the history of industrial art has this lack of contact between artist and industrialist been so unfortunate, for the contemporary designer is qualified for active partnership with industry not only by his ability but by his lucidity of purpose and because his mind is attuned to the rhythms and is responsible to the stimulation of the machine age. He could give an unforgettable splendour of direction to industrial art, and could be as great a master of the new materials and processes as mediaeval masons were masters of stone and of the tools with which they shaped it.

But only a few individuals who have the courage of their own good taste employ designers intelligently. An illustration of the attitude of mind that still exists in many industries that could derive economic advantages by collaborating with designers is to be found in a report prepared in 1928 by two of H.M. Inspectors of Schools, to assist the Joint Standing Committee (Industry and Education) of the British Cotton Industry Research Association in discussing the training of designers for printed and woven fabrics. Entitled, *Design and the Cotton Industry*,[1] it surveyed with elaborate care the methods customary in the cotton industry for

[1] A Report of H.M. Inspectors on existing conditions in the Industry and the Schools. London: H.M. Stationery Office.

obtaining fresh designs. It was a revealing example of traditional industrial patronage, and suggested that the ideas about design recorded by Mr. Samuel Sidney in the middle years of the last century were still flourishing in post-war Lancashire. It indicated that the cotton industry was only driven to consider education in design at all when the prosperity of Lancashire began to totter. "Compelled by intensified foreign competition and the consequent loss of markets for plain goods, manufacturers are turning their attention to the production of fancy goods in which design is all-important." A comparable situation in another branch of industry was described by Mr. Sidney who also mentioned the economic consequences of making "fancy goods" without design.[1] "THE GILT TOY AND MOCK JEWELLERY TRADE, once one of the staple employments of Birmingham artisans, has dwindled away until it now occupies a very insignificant place in the Directory. Bad cheap articles, with neglect of novelty and taste in design, ruined it. In cheap rubbish foreigners can always beat us, but the Birmingham gilt toy men made things 'to sell' until no one would buy."

Design and the Cotton Industry stated that economic circumstance rendered it "vitally important that effective means shall be found to draw into the service of the industry men and women of trained artistic ability." As this was being done by other industries, "the cotton industry cannot afford to remain indifferent." The report admitted that even if the art schools encouraged students to consider the prospects

[1] *Railway Rides*, p. 103.

of the industry, "whatever be the attitude of the authorities of the schools, the well-informed and capable students will, as far as they can, choose that branch of applied art which offers the greatest opportunities of advancement, professional, social and financial." In the paragraph following this admission we learnt that in the cotton industry "changes, even if they are recognized as improvements, must be based upon current practice, and they will be slow, piecemeal and, usually, imposed by economic forces." Presumably the views of the industry on the value of art and the remuneration and status of designers will be subjected to this piecemeal process of change: £6 a week is considered adequate for first-class designers at present. After mentioning that particular valuation, the report naively recorded the fact that: "The supply of finishers and ordinary designers is said to be satisfactory, but really good designers are rare."

The sources of supply for designs for printed and woven fabrics were described in detail, also the manner in which those designs are inspired. Whether art and industry could enjoy partnership in this field may best be judged by the following quotations from the report:

"Fabric painters obtain their designs from one or more of the following sources: works studio, English commercial studios (chiefly in Manchester), free-lance designers (including, occasionally, art students), and French commercial studios."

"The designs may be either entirely new and the expression of the designer's own originality, or, more often, the result of suggestions made by clients and based upon their opinion of what is likely to sell in the particular market for which they are catering. These opinions are, of course, usually based upon designs which have already proved successful."

"It does not appear to be unfair to say that the designer in a studio is generally looked upon as a person of no great importance."

The British Institute of Industrial Art estimated that only 3 per cent of the designs were bought from free-lance artists. When the free-lance "presents his designs for inspection, the treatment he receives sometimes reflects too crudely the subordinate status of the designer in the studio, and he retires from the effort in disgust." Incidentally those firms employing free-lance designers "have reaped their reward in the freshness and originality of their products." There were occasional attempts to use a French designer on the spot, but it was found that "if he is transplanted to England and particularly to Manchester he loses his freshness in a short time."

French designs are bought by managers, heads of departments or salesmen who, without any training in design, "have formulated piactical standards of taste by long experience in handling fabrics, by critical attention to designs and by close contact with markets." It is suggested that these buyers object to innovations. It is almost an English proverb that everything naughty, new and nice comes from France, but it is uncomplimentary to English discernment that there should be "evidence that the free-lance artist is more likely to get his designs placed if he commissions a

French agent than if he tries to place them himself." Also "it is said that the French designers have a certain contempt for the English buyers of designs and do not show them their best products. It is difficult to see the reason for this and in Paris it is stated without qualification that the best designs come to the English market."

Manufacturers of woven fabrics got designs from "public designers" (whatever they may be) and from the design rooms of their own mills. The same futile muddling with design was exposed in this branch of the business, and the whole report showed the ignorant contempt in which art is held by the bulk of the firms engaged in the cotton industry. In the words of the compilers of the report: "one is led to wonder whether the more frequent employment of professional artist designers of high standing in England, and a closer connection with the schools, might not do something to counterpoise the weight of French tradition and psychological aptitude."

But the cotton industry is not exceptional. Anybody who inspects in detail the weary miles of indifferent rubbish that line the aisles of the British Industries Fair every year; who examines the fancy goods departments of retail establishments, large and small, who mourns the snail's progress of decent design in the trades that are responsible for making the Englishman's home what it usually is—any such conscientious observer must be overcome, not only by depression, but by curiosity regarding the cause of this queer stagnation in our industrial life.

There is something lacking. Something which other countries possess and exploit and which enables European and American goods to find and to keep markets that are beyond our reach, for until recently we have not realized that France, Germany, Sweden and Austria have what can only be described as an anonymous export, not an invisible export, for the presence of invisible exports is known, while the existence of this anonymous export is still unrecognized by British Industry, although the Board of Trade has identified it. Industrial administrators, manufacturers and large numbers of well-educated and responsible people who are incapable of recognizing this anonymous export or of defining its character, will admit that things of metal and glass and pottery and wood that are designed abroad look different from things of the same type that are British made. They do look different, and in many ways they look better, because in most European countries far more things are well designed: in Britain they are only well-made, for we have been off the design standard for nearly a hundred years, and unless we get on to the design standard again, our prestige in overseas markets will, as the world grows more enlightened and artistically expectant, decrease to the vanishing point. Good design is the anonymous export, and it is the foundation of an international structure of propaganda about the superiority of European and American goods.

Organized propaganda concerning national ability in design is not a new idea; it enjoys to-day fresh modes of expression, of which the American film is perhaps the most powerful; but in France it began in the days of Colbert.

From that time France has led the world in artistic fashions. Colbert's policy of founding an effective alliance between design and industry is the reason why a Lancashire cotton manufacturer is prepared to buy the work of an English free-lance designer if it is sold to him through a French agent: France has an unshakeable reputation for design.

New movements in design are crystallized at the quarter-century international exhibitions that are held in Paris. Magazines all over the world illustrate French furniture, French interior decoration and textiles and glass; and other countries are learning to advertise their abilities in design by the same methods. Within eight years Sweden held two exhibitions on a big scale, at Gothenburg in 1923 and in Stockholm in 1930; exhibitions that have been talked about and illustrated everywhere, and which evoked the admiration of this country in particular by the excellence of their lay-out and the standards of design illustrated by the exhibits. In the Spring of 1931 Swedish design invaded London in a small, highly selective exhibition of contemporary industrial and decorative work. In the Spring of 1934 there was an Austrian Exhibition in London of the same kind. Everywhere photographs are being published of Swedish, German and Austrian work in those branches of industry that should enjoy an inventive alliance with art, while Hollywood sends her film stars driving into every town and city on earth in the most shapely cars.

Creative original design does not greatly interest contemporary England; but it does interest the rest of the world, and as a nation we do little to satisfy that interest. The manufacturer may be able to hold the home market by producing articles that are appropriate for the people who make their homes in near-Tudor houses that look like cuckoo-clocks, but unless he can find and exploit laɪge, new barbarian markets abroad, which would find flashy rubbish acceptable, he will be limited to the home market. And in time even the home market may be stimulated to rebel.

If industry is to move forward, its controllers must bring into the factory the abundance of designing talent that exists in this country. Collaboration between design and industry must be deliberately planned. And design, which is too often regarded as either a nuisance or an extravagance, must be honoured as an economic factor. It is hopeless to expect any political party or fusion of political parties to produce a plan for the association of design and industry, as Colbert produced a plan. Politicians merely apply faith to a problem, and political faith is usually a composition of funk, watery ideals and greed. No problem is judged on its merits, and although politicians may barely acknowledge the existence of this particular problem by setting up a committee, the recommendations of a committee are always apt to get shelved unless they can be exploited to drag a government out of the mire of a crisis. The Trades Unions, which might have been expected to take an alert interest in the things that were made by their members, and in education that would improve the character of industrial products, have never risen above the bitterness of battles over wages and hours. What work is about or the way it is done are alien matters that cannot claim the time of Union officials; quantity not quality interests them, how much cash for how little time is the theme that engages their earnest attention, although they can occasionally spare some of their energy for diverting the public with exhibitions of their dignity, when for example they withheld official approval from the election of a little girl as Railway Queen in a beauty contest, because her father although a railwayman was not a member of the Union. In the hands of clerks without vision the Unions can never produce a creative plan, for they abhor the essence of all industrial planning, which is partnership. Industry must help itself.

But the gravest problem of patronage for industrial design is the incompatability not of the designer and the manufacturer, but of the reactionary distributor and the progressive manufacturer. The progressive distributor quickly discovers the progressive manufacturer and ignores the rest, and gets most of his goods abroad. The situation is something like this:

THE RETAIL BUYER—OLD STYLE

He began the job when people bought what the shop-keepers chose to give them, without complaining. It was a nice, passive shut-eyed public in those days;

it wanted things homely and comfortable and durable, and the poorer parts of it were quite happy with imitations of things that were homely and comfortable and durable. None of this art tosh: people who wanted that went and hunted in the rubbish heaps of the old curiosity shops.

But that habit spread, and people got to wanting old shapes and patterns; touch of the antique in their chairs and something a bit chintzey in the curtains, so he gave 'em good reproduction stuff at all prices, from high-class down to medium-class goods. They knew where they were with the old things, and so did the makers, and if you wanted a bit of variety, well, you could always order something original, a Jaco hall-stand in mahogany instead of oak, or a coffin stool in walnut, or you could let yourself out with new colours in art pottery.

Of course there were little bits of fashion, but what went well in one year was generally good, with a few variations, for three or four years. The public wants the old English fireside comfort, and plenty of cushions, and they like things smartened up with a bit of orna-ment. That's why Jacobean was always more popular than the Louis style: too foreign, that was, rich, of course, but the English home likes its richness solid. A little of what you're used to does you good. Every-one starts keeping house with something they've got from Auntie or Dad, and they want something like they've started with. He knows what the public wants, and he never shows 'em anything else.

THE MANUFACTURER—NEW STYLE

He began to realize after the war that British Industry
can't live by craftsmanship alone. To the reserves he
set aside for the experiments that maintained his
technical efficiency as a producer, he began to add
odd sums for even odder experiments, fees to experts
in design for criticizing his products, fees at last to
designers. He is as willing to experiment in design as
he is to experiment in new technical processes, but the
retail buyer says: "Don't go and make things difficult,
old man; *we're* in touch with the public and we *know*
what'll go and what won't—we couldn't possibly take
more than a dozen, and that 'ud be a risk."
He doesn't see quite why he should be expected to
carry the baby all on his own, but he can't ruin his
business by going direct to the public, so he earmarks
a little of the budget for these experiments, does some
high-powered selling himself on a few retailers, because
he believes that there *is* a market for the new stuff (it
may be pottery, textiles, furniture, glass or whatnot),
and perhaps just covers the cost of his experiments.
And the old men on his board use the most blight-
ing imprecations regarding his activities, such as:
"Idealist!" Every year it is more difficult to get those
sums passed by the board. He is looking ahead, and
knows that they are vital injections for the refreshment
of the firm's commercial life, but he is obstructed by
the accountant. Unfortunately none of his new stuff
helps to sell the old-established stuff he makes: his

experimental designs for twentieth-century taste and the customary products of his factory belong to wholly incompatible civilizations.

THE RETAIL BUYER—NEW STYLE

Ever since the war he has been trying to make people feel that they haven't got to go to Paris or Berlin or Vienna for every new idea. But when he goes abroad and gets well-designed things for his enterprising London shops his customers say: "But everything's foreign in here—it's a scandal that you don't sell British goods." He would if he could get them; but the manufacturers say, "Well, if you'll guarantee to take 1,000 we don't mind, just this once, putting down the money for the experiment!" And when the experiments are made, often enough the results are far more expensive than imported things of the same kind, in spite of tariffs.

He wants to give people a feeling that there's always something new and exciting to be found in his shop, and there is, but the bulk of it isn't British made. Things are brought over from the Continent—glass, pottery, light furniture, fabrics, fancy goods; and English manufacturers who can't or won't design are set the unseemly job of copying them. Meanwhile they go on offering the old, old goods; the stuff their fathers made so proudly and sold so easily; and it looks as dowdy and incongruous in a really modern shop surrounded by things of twentieth-century design, as a 1900 motor-car looks in present-day traffic. He doesn't pretend to exact knowledge of what the public wants; but he does know they'll always come to see something new, no matter how unusual, no matter what its price.

THE MANUFACTURER—OLD STYLE

Ever since the war he has been hoping that pre-war conditions will return. All this modern naked stuff, bare walls, bare rooms, nudist movements—they're all symptoms of social upheaval. Presently people will settle back to their normal ways and want stuff that has some skill and ornament about it; anyway, most people want it still, but they can't afford to buy owing to the insane wages Labour has to be paid; can't produce the stuff cheaply if there's to be any profit. All this nonsense about new design: why, it means laying down thousands in new plant, spending good money to give people gim-crack made-in-Germany-looking stuff. It's un-English. Worse. It's unpractical.

The firm was founded by his great-grandfather, and it's grown to its present size by making things solid, not making 'em flashy. What's the good of some of these larky young buyers showing him glittering affairs from some Viennese bazaar and saying: "Why don't you make something like that?" What his father made and sold thousands of is good enough for him, and when the world gets back to normal then people'll be wanting the same good old solid stuff again. People, the British public, don't change, though the world gets out of joint now and then. Designers——? Well, young Alf (though he's nearly sixty now) who began in the pattern shop when he was a lad knows what we want, and he's always been handy with a pencil— got an eye for colour too. We don't want any art in our business.

*   *   *   *

In spite of those arrogant obstructionists, the reactionary retail buyers who insist that they know what the public wants, the public occasionally wants something that somebody with enterprise has had the courage to put before them. There are probably more men and women with better taste and saner judgment in the shape and colour and fitness of the things that go into their homes than this nation of shopkeepers suspects. When people begin to apply to the purchase of china, glass, furniture and so forth, the critical standards that they apply when buying their clothes and their motor-cars, industrial art in England will get more active encouragement than any number of exhibitions and sermons can supply—it will get the only effective encouragement possible in the present economic organization of the world : demand by consumers.

CHAPTER IV

EXAMPLES OF DEVELOPMENT IN DESIGN

> "It is only a secondary aim of the designer of
> machinery, if his machine is to be looked upon
> as a thing delightful to be seen—delightful as a
> thing harmonious in its physical shape—though
> a good machine may always be so looked at and
> is always a thing of beauty in the sense that the
> beautiful thing is that which pleases when seen."
>
> *Art and a Changing Civilization*,
> by ERIC GILL

> " 'I know that some excellent persons have been
> writing lately about the beauty of a swift-
> gliding motor-car and things of that kind. They
> are right, in one sense of the word. For there
> is a beauty in mechanical fitness which no art
> can enhance. But it is not the beauty of which I
> spoke.'
> " 'And therefore,' observed the bishop,' we
> ought to have another word for it.' "
>
> *South Wind*,
> by NORMAN DOUGLAS

WHEN any new utensil or machine is invented,
it passes through several phases of development
which are controlled partly by mechanical improvements which may modify the spatial demands of the
object and may render it more compact and convenient
in use, and partly by fashion in shapes. Before the
commercial machine age, evolution in design was
usually a graceful process. New forms would be boldly
invented without being cramped by a sense of obliga-

113

tion to some prototype. A good example of the vigorous and healthy growth of design is afforded by the history of clocks during the seventeenth century. The clock began in England about the middle of that century as a small piece of mechanism accommodated upon a bracket with its weights dangling down on chains. In this early stage the "lantern" or ":bird-cage" brass-faced chamber clock had an important part of its mechanism exposed, rather untidily. The next stage was the enclosing of the metalwork by a hood, so that the hanging clock exposed a metal face and a wooden case; but the weights still hung down, and so

FIG. 14—*A seventeenth-century lantern or bracket clock, first stage, with mechanism partly exposed*

FIG. 15—*The mechanism completely enclosed by a wooden hood, but the weights still exposed*

did the pendulum, when that device was introduced. The next advance in clock design was to provide a covering for the weights and the pendulum, a case which would also support the clock itself, and so the long-case or "grandfather" clock arrived, with its form influenced only by contemporary architectural taste which prescribed appropriate proportions and certain moulded details.

As the mechanism of clocks improved, it was possible to accommodate all the machinery within a relatively small space, and the case was dominated by the dial which had been evolved by purely functional considerations. Continental extravagances during the eigh-

FIG. 16 — *Works and weights completely enclosed: the grandfather clock of the late seventeenth and eighteenth centuries*

FIG. 17—*The complete disappearance of mechanism: an electric clock face plugged into a wall*

teenth century provoked various ornate imitations, and ultimately produced in England the sepulchral timepiece of black marble with classic columns and nondescript figures that in the eighteenth century would have certainly been cupids but in the nineteenth were definitely angelic. We have to-day reached the logical end of clock design with the electric clock. The mechanism has almost disappeared. All we need is a dial, hands, and a plug to connect with the electrical supply. But we are so bullied by prototypes, that many of these compact and beautifully simple clocks are surrounded with a perfectly useless and quite empty case. It may take half a century to lay the ghost of clockwork.

Examples of designs which have developed in our own machine age are gramophones and wireless receiving sets. We begin with the phonograph.

This is a piece of exposed machinery mounted on a

FIG. 18—*Machinery naked and unashamed: the early phonograph, a frank piece of engineering*

simple base. There is no attempt to disguise the fact that it is a machine. Its functions are externally visible. It is mechanism in the raw.

FIG. 19—*The early gramophone with a little injudicious gaiety in the trumpet*

The next stage is the early gramophone. Here again mechanism is naked, but not quite unashamed. The mechanical hardihood of the design is softened by an attempt to make the trumpet decorative.

In the next stage the trumpet disappears, and the gramophone turns into a piece of furniture with the horn concealed inside. It becomes a cabinet, and as such is susceptible to all the traditional influences which could be borrowed and pressed into the service of external decoration.

FIG. 20—*When the trumpet was accommodated in a cabinet, gramophones and radio receiving sets passed through a phase of disguise, and those who made cabinets for their accommodation borrowed "freely" from England's glorious heritage of furniture design*

Now the radio-receiving set began in precisely the same way as the phonograph and the gramophone. At first it was a piece of complicated and excessively untidy machinery. The crystal sets of the early 'twenties with their tangle of cables and ear-phones and apparatus, exposed their entrails to the world with the unreticent brutality of a typewriter. Then came refinements of mechanical invention; valves and loud-speakers and all manner of subtleties, and from an engineering point of view, compactness became possible.

The gramophone and the radio set were presently combined and the radio-gramophone as a composite piece of apparatus displayed such a violent reaction from the crude mechanical exposures of earlier days that it was sometimes difficult to tell, when all the doors were closed, exactly what this new piece of furniture was.

For instance it might borrow its inspiration from early Stuart or Carolean furniture. It might hark back even further, and an imitation Tudor food hutch might house the latest mechanical triumph of the twentieth century.

FIG. 21—*Although the radio cabinet and the gramophone itself could have been distinctive pieces of twentieth-century furniture, the disguise phase bound them for a time to traditional forms*

Again it might seek a borrowed elegance from the William and Mary period, elevating upon legs a cabinet that contained all the essential mechanism.

FIG. 22—*But the gramophone did attain a compact acknowledgement of its function. (An H.M.V. model)*

But at last there was a rationalizing process, and the work of engineers was no longer accommodated in cabinets that were earnestly pretending to be something antique.

The radio-gramophone was acknowledged as a

FIG. 23—*And the radio gramo-phone also attained a compact acknowledgement of its function*

piece of furniture with a function and form of its own. The wireless receiving set as its apparatus became more compact and made fewer claims on space, began to acknowledge its obligations to truth about its function, and in that acknowledgement it could display simplicity and comeliness, and could provide the cabinet-maker with the opportunity of using decorative veneers.

The final form of radio apparatus may be a small glass and metal case, taking up very little space and merely embellished with the studs and dials needed for adjustment. Like the clock, however, it may suffer from a perpetuation of needless bulk, because people

FIG. 24—*The radio receiving set acquired concise and lucid expression*

FIG. 25—*Thus functional fitness began to dominate its form. (An Ekco Receiver, Model 65. Cabinet designed by Wells Coates)*

have got used to the idea of a certain amount of space being taken up by a wireless set.

It will be appreciated that our habits of thought about shapes determine upon quite a large scale the external character of the things produced by the

123

abundant mechanical inventiveness which distinguishes this age. These habits of mind may limit the inventiveness by the demands they inspire. The march of invention in artificial lighting illustrates this form of limitation.

In the artificial illumination of rooms design *should* have been influenced by the character of the illumina-

FIG. 26—*The chandelier designed to hold wax candles*

tion, and it was in its initial stages when candles were used. The candlestick, the wall sconce and the candelabra were all functionally designed to provide the maximum of illumination and the maximum of protection from dripping wax. Throughout the seventeenth and eighteenth centuries the form of candelabra was controlled by these functional needs. The embellishment of the necessary structure graciously accorded with contemporary fashions. Experiments were made with various materials and the crystal chandelier with

FIG. 27—*The crystal chandelier which enhanced the whole character of candle light with wavering reflections*

its lustres and myriad reflections represented the peak of achievement in design for this type of illumination.

When a totally different form of lighting was invented, namely gas, the new power was adapted to forms which resembled very closely those employed for candles. The gasolier was an adaptation of the chandelier.

And when the final convenience of electric light

125

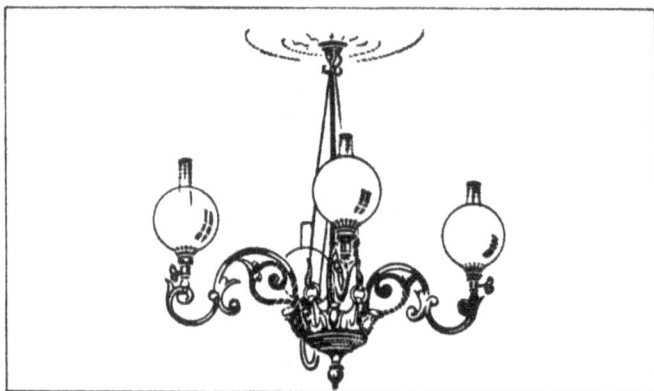

FIG. 28—*Gas light borrows the form designed expressly for candles.
The Gasolier*

FIG. 29—*Electric light borrows the form originally designed for wax
candles, but inverts the arms so that the lights hang. The Electrolier.
(Early twentieth century)*

began to compete with gas the electrolier again followed the candle-holding forerunner, even though the

electric bulb soon outgrew its early feebleness which had justified a Cockney's description of it as "a red-'ot 'airpin in a bottle!"

Perhaps the most obtuse rejection of the possibilities of electricity was the imitation candle with an electric bulb on the top of it. True, this dummy candle enabled antique candelabra to be used, but the true function of electric light was not immediately appreciated when it became a practical possibility. Candles and gas had implanted the idea that the source of light must always be visible. A naked flame or a glowing filament were familiar. The convenience of electric light was appreciated at first for its obvious economies of effort.

FIG. 30—*The logical development of electric light: diffused light with its source invisible*

No matches had to be struck and carried waveringly across the uncharted gloom where furniture in the wrong place might entrap the unwary.

Glaring light was brought into the room and then shaded. The glare was considered essential, whereas the real solution of illumination by electricity is light from an invisible source with conveniently disposed apparatus for concentrating light on any particular spot, such as a desk light which gives an area of illumination without revealing the bulb at all.

Although concealed lighting is the logical development of design in illumination, the clean, simple, dust-proof functional fittings of the present day, do r··· resent an extremely able use of materials and a˙ compact and agreeable form of lighting.

Two examples of development in machine design may be given which supplement those already illustrated in Chapter I. The electric tram represented a complete break with tradition. It might have taken the form of a railway carriage which had originally been based upon the form of the stage coach. Actually the early trams with their open upper decks were fairly simple.

For example the type shown in Fig. 31 was in service at the beginning of the century and was a plain expression of purpose.

A little ornament was permitted to creep in and the railings on the upper deck were rather self-consciously decorative. The stairs to the upper deck took up a lot of room and the driver and the conductor were thrust firmly into the fresh air (and the bad

FIG. 31—*The electric tram in 1900*

FIG. 32—*The latest type of electric tram*

weather) and roofing space over their platforms which might have provided extra seating accommodation on the upper deck was dispensed with.

The latest example of tram design is wonderfully orderly and lucid in its form. It is completely tidy. It is luxuriously comfortable and all the loose ends and oddments of apparatus that disturbed the outlines

FIG. 33—*The horse-bus*

of the earlier model have been skilfully accommodated and smooth, clean lines are the result.

FIG. 34—*The early motor-bus*

Motor-bus design was for a long time under the spell of the horse-bus.

The early motor-buses looked like horse-buses clumsily dumped on to a chassis.

It was not until the London General Omnibus

FIG. 35—*The "B" type motor-bus put on the streets in 1910 by the London General Omnibus Company*

Company put on the road the "B" type of 'bus that the design of 'buses was really adjusted to the needs of higher speed and the comfort and protection of drivers as well as the comfort of passengers.

The post-war experiments of the London General Omnibus Company in 'buses are, perhaps, the most brilliant example of progress in machine design that any organization can afford.

The "K" type of 'bus which appeared on the London streets in the early 'twenties was soon replaced by the "NS" type, in turn to be supplanted by the six-wheel

131

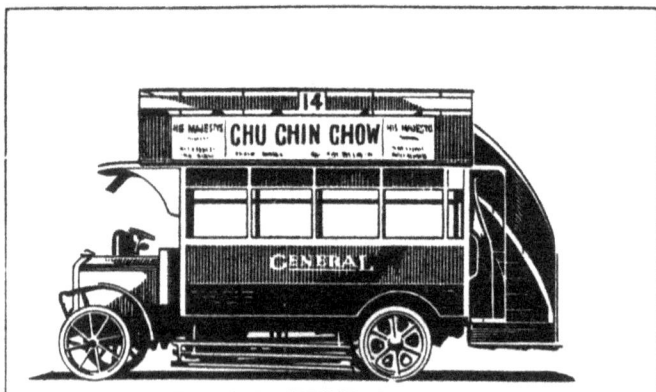

FIG. 36—*The "K" type General Omnibus of the early 1920's*

FIG. 37—*The next development in 'bus design: the "NS" type of General*

type, and these have given place to a four-wheel stream-lined type with an enclosed cabin for the driver and the most comfortable seating for the passengers and an internal staircase for reaching the upper deck.

The final form of a 'bus and motor-coach may be

FIG. 38—*The contemporary motor-bus with driver's cabin protected by glass, covered top, and internal staircase, and deeply sprung upholstered seats.*

FIG. 39—*The motor-coach and the motor-bus without a bonnet and with the engine behind the driver have arrived. The "ghost of the horse" is laid for ever by this type*

133

the type which dispenses completely with the bonnet and so at last lays "the ghost of the horse" which has hitherto run before every motor-vehicle.

Industrial design although it may be hampered by a prototype may sometimes employ material which makes it impossible for any respect to be paid to traditional forms. An instance of this is the comparatively recent appearance of furniture made of manipulated metal tubing. The chair form possibly reached the zenith of its comfortable and decorative adjustment to the needs of the human body in the early

FIG. 40—*Functionalism in wood: the Windsor or stick-back chair*   FIG. 41—*Functionalism with manipulated metal tubing*

years of the eighteenth century. The Windsor or stick back chair was an entirely functional solution in wood of the problem of seating. The modern chair with a tubular steel or copper frame with a slung seat and back is a perfectly functional solution to the problem of seating with a less friendly material than wood.

Designers of this new metal furniture had to develop a technique of their own because their material demanded it. As a later chapter describes, the disguising of metal so that it looks like wood does not present any difficulties to the manufacturer of furniture; but some materials impose functional frankness by their nature, and metal tubing is one of them.

The examples quoted and illustrated in this chapter suggest that when industry produces something that has no prototype, then after it has passed out of its purely mechanical stage (if its character is mechanical) disguise is sought for it. With machine design the absence of a prototype confers complete freedom upon the engineer who generally uses it wisely unless he is entrapped into making some decorative "additions" to accommodate an imagined need for "ornament." (Nothing is quite so painful as an engineer trying to be artistic.) When machine design starts with a prototype we have seen that it may take many years to outgrow the limitations that respect for the ancestral form has imposed.

In industrial design inventiveness is liberated when strikingly new materials are brought to the service of old needs, and designers are by the very nature of those materials released from the repressive necessity of imitating an old form.

In a well-ordered, well-educated world there would be no borrowing of traditional forms. Designers would study tradition for what it could teach them about proportion and about the skilful adjustment of means to needs with various materials; but they would solve every problem of design on its own merits with the materials most appropriate to the solution. That, as the last chapter suggested, is what the modern movement in design is attempting.

CHAPTER V

EXAMPLE OF DEVELOPMENT OF DESIGN IN MATERIALS

WHEN industry has to provide materials which will be used by a skilled and critical designer there is greater incentive to the manufacturer to embark upon innovations. His experiments will not be turned down by a repressive distributor, blind in his belief that "he knows what the public wants." Perhaps manufacturers of building materials are given the greatest opportunity for experiment because a proportion of their products is used by trained and educated designers—by members of the only profession that is trained to design, the architectural profession.

The manufacture of glass illustrates, perhaps more lucidly than any other section of industry, the effect of imagination upon the nature of a product. In the making of structural and decorative glass there is in this country a technical fecundity that has changed the whole character of glass as a material for building. This inventiveness is not sporadic; it is continuous, and it is cautious. In this branch of industry designers are often present at the councils of manufacturers. There are more working partnerships between technicians and designers than are suspected, and some of the great English glass-makers are demonstrating the practical value of such collaboration.

Of most English manufacturers it may be said that

they will never release for consumption any product that is still in its experimental stage. Their technical integrity, if it may be put that way, is unshakeable. None of the revolutionary types of glass that has appeared during the last few years has been released casually when it was still in the bright young idea stage; none of them has been allowed near an architect's office until it has thoroughly satisfied its makers.

Structural glass has developed so many surprising qualities that it can never again be regarded merely as a window filler. It has been suggested earlier that some English architects are still reluctant to admit the implications of the structural revolution, and willingness to do so is unfortunately curbed by archaic building regulations, and by those anonymous and regrettable arbiters of architectural form—the ground landlords of our towns and cities. The opportunities for experimenting in new techniques of design, in which the changed character of contemporary glass could be shown, are limited; but one of the limitations is what may be called the window complex. Our chief traditional precedents for the use of glass are windows, and windows of limited size at that. We fail to regard glass as a substance of diverse possibilities. We think of it as a substance for a set and particular purpose, for the admission of light in association with materials that will guard its natural fragility, such as wooden or metal glazing bars.

Technically, glass has outgrown many of its traditional limitations. Reinforced glass is made, with wire-netting buried in its core, a material with new powers of resistance, something outside the window tradition, and when it acquired the refinement of a square mesh (the "Georgian" type), with the intersections of the wire electrically welded, it became a reinforced glass that was also intrinsically decorative. This is one rather obvious example of emancipation from the hole-in-the-wall stage. Reinforced glass now comes into the category of materials from which the wall itself can be constructed. It is also an instance of industry making a contribution to the possibilities of architectural design, which, like the invention of lifts and reinforced concrete, could change the structural character of architecture.

The idea of glass walls for industrial buildings, for clinics and schools develops logically with the structural revolution; and the invention of a glass that admits a high proportion of natural ultra-violet radiation is an excellent reason for expanding window space to the maximum area. (Any doubts that may have existed about the permanent ultra-violet admission powers of this type of glass were finally dispelled by the reports of tests extending over some years which have been recently issued by the National Physical Laboratory.) But a complete reversal of all conventional and accepted notions of what glass can and cannot do is brought about by "Armourplate" glass. This is plate-glass subjected after manufacture to a toughening process, so that it will bend beneath a weight until the weight is increased beyond the supporting power of the glass, then it disintegrates and breaks up into powdery crystals. This glass can bend and twist under pressure, and it has great heat-resisting powers. The claim of glass to be an independent and ubiquitous building material is now indisputable. This last technical advance should dissipate the window complex, and should stimulate the creation of a new technique of design.

Memories of Continental exhibitions, of strident stunts, and colourful Gallic impetuosity, often mar the beautiful directness that English designers can achieve when they are working with unconventional materials. Glass in decoration is still an unconventional material: it is usually employed for mirrors or trimmings: it is only just beginning to be used inventively and with subtlety in conjunction with lighting, and we are only just gaining full consciousness of its significance as texture.

The ornamental types that are made, naturally include some archaic curiosities, but artists of the calibre of Paul Nash have occasionally been commissioned to design new patterns, and there is no need to go beyond England to find an abundance of excellent designs in decorative glass. That is not an empty "Buy British" statement: its truth can be tested by examining the existing varieties of English ornamental glass.

Less heartening are the statements that can be made about the totally different branch of English manufacture which is responsible for domestic glass. Glass that is often beautiful in shape is tortured with ornament. Design still dies the death of a thousand cuts. Good shapes were retained throughout a large part of the nineteenth century, but their beauty was obscured by unrestrained brilliant cutting. The heavy cut-glass tumblers, wine glasses and decanters of mid-Victorian times often mask by their glittering ornamental complexity the most agreeable lines. This affords an example of hand-craft in alliance with industry losing all contact with design, and coming under the control of uneducated pattern-makers. Occasionally some firm has the good fortune to find a designer like Keith Murray and the intelligence to let him try experiments; but there are only a few isolated instances of such productive collaboration. Generally the only alternative to the mark of the mid-Victorian beast is wholly undistinguished, something plain and blameless and rather heavy-handed.

The cocktail has a lot to answer for: much spurious modernism has been inspired by it, and most of us have shuddered as we have imbibed from glasses designed in a spirit of squalid gaiety.

That there should be any sort of achievement in the design of drinking-glasses, decanters, bowls and vases is remarkable. For so long have English manufacturers with a very few honourable exceptions rested on their doubtful and dusty laurels, that it is difficult to persuade them that the taste of the world is changing. They have had no structural revolution to stimulate them; they are not consciously taking part in any modern movement of thought, design or industry. Yet the examples of well-designed glass that are made in England to-day suggest that great abilities are dormant; but at present the domestic branch of the industry is half a century behind its virile and progressive elder brother.

CHAPTER VI

EXAMPLE OF DEVELOPMENT OF DESIGN IN EQUIPMENT

A PART from factories, the only type of building in which decorative disguise might easily have been abandoned was the office. Surely this was the place where industrial design could have developed the functional certitudes that were repressed in so many other directions. But although every new invention in structural technique, every revolutionary improvement in heating and lighting and ventilating, every decrease in the need for mobility and effort occasioned by such inventions as the telephone and the dictaphone, challenged the idea of disguising purpose, the idea remained embedded in the minds of those who chose or designed office equipment. In a few centuries, the wealth, power and glory of commerce has changed the office from a poky little hovel in the corner of some warehouse, into a spacious, and, so far as the board room is concerned, majestic place.

Up to the end of the nineteenth century the dreariest tradition clung to the office. The offices which Dickens described were always grim. "The door of Scrooge's counting-house was open that he might keep his eye upon his clerk, who in a dismal little cell beyond, a sort of tank, was copying letters." Scrooge was not a model employer, but even the benevolent brothers Cheeryble had a counting-house which was a fairly near approach to a prison cell if the illustration of it by Phiz is a true interpretation of the text. The savage disregard for human convenience and comfort which has frequently distinguished the organization of money-making in the past has been dispelled, not, as a moralist would wish, because of a change of heart, but because of an increase in sense. The bottom rungs of the ladder need no longer be uncomfortably submerged in gloom and foul air.

It was in offices in cities that the worst traditions prevailed. Offices attached to factories in the growing industrial centres of the nineteenth century had more space, and although the worst was often made of it, at least the likeness to cramped dungeons was avoided. The attitude that the average Victorian and Edwardian householder had towards the domestic staff was shared by the average employer. Any place was good enough for them to work and live in, and the employer was fortunately absolved from bothering about where the brutes slept.

The private office of the employer himself used to reflect the ponderous standards of comfort that controlled his domestic taste in furniture. Although during the last third of a century. the employees' section of the office has shaken off all its dreary and depressing characteristics, the board room and the private offices of directors still maintain an inappropriate relationship to the domestic interior. Up and down England there must be hundreds of board rooms that are furnished as Georgian dining-rooms or worse—Jacobean dining-rooms. As a setting for modern business practice, a mahogany table and chairs of neo-Chippendale type are hardly suitable: the association of such furniture with the art of conversation as it was understood in the eighteenth century, and, when the ladies had withdrawn, with the felicitous story-telling of three- or four-bottle men, render it incongruous for the bleak platitudes of the board meeting, the resolutions "to proceed upon the accustomed lines" and the reiteration of that potent assertive, "quite." But the board room can seldom reflect the progress that has been made in the design and planning of offices, for its decoration, furnishing and equipment depend upon the personal taste of the magnates who use it and unless they employ an intelligent architect they will merely get the wrong sort of dining-room.

But we may glance at the advantages the office has derived from the materials and services of industry.

The character of office design has been changed by electric light, the telephone, central heating .and a host of new materials with special properties. Even now, the services which are available for equipping the modern office are not always considered in the initial stages of planning. The position of the telephones, both the G.P.O. instruments and house telephones, are scttled nearly always after floor spaces have been planned out with partitions into various groups of offices. How many offices are there without trailing wires connecting instruments with tables, which require the most expert footwork on the part of visitors, unless when leaving they drag from some impressive directorial table the house and G.P.O. telephones, and if they are really unlucky the desk lamp too, having failed to avoid entanglement with those crawling cables? In the U.S.A. there is a system which provides for wiring every floor of a new building with power, light and telephone cables. Junction boxes are arranged at close intervals all over the floor before the blocks or floor boards are put down. This system allows a desk or a row of desks to be placed anywhere, for light and telephones to serve them can easily be picked up, also power to operate radiators, adding machines, and other mechanical devices without running surface cables. This simple method is applicable to the two principal types of office planning: the large open space with desks and tables conveniently arranged all over it, or the warren of private offices created by innumerable partitions.

The introduction of the telephone has reduced the need of mobility in an office, and has probably cut down the number of messengers employed in any organization by at least 30 per cent. It is difficult now to imagine how business could be conducted without telephones. A horribly insanitary affair was used in pre-telephone days, namely the speaking tube. You blew down it forcibly and a whistle sounded as long as you kept blowing or until somebody uncorked the other end. There was always a disconcerting doubt as to when it was your turn to speak, and master and man frequently blew at each other in an unhygienic and exasperating manner. But the telephone has now developed refinements which have still further decreased the need for mobility in the office. The managing

145

director with the appropriate instrument sitting at the desk in front of him can conduct a board meeting without his co-directors leaving their individual sanctums. His words are broadcast to them by an amplifier, and they can reply when they think it is their turn. No doubt television will contribute other facilities to directorial intercourse, and Captains of Commerce may sit in their country seats and hold a succession of board meetings in London, Paris, New York, Manchester, Sydney and Cape Town in the course of half a day.

The two main systems for planning offices just referred to are based upon the desire to attain,

1. Complete visibility.
2. Privacy.

The open room with desks dotted all over it secures visibility. A very large space in the middle of that room with a managerial desk in the centre of it secures complete privacy for conversations between the manager and anybody he wishes to talk to.

Again, visibility is achieved with partitions glazed above dado height, which provide individual offices, and enable the activities of the staff to be under observation. The method for which most English firms display an affection is the individual office formed by partitions of some opaque material. This complicates the problem of ventilation, and unless the partitions deaden sound, many indiscreet utterances may reverberate throughout a whole group of offices. (The writer recalls a large organization, in which a directorial voice could often be heard thundering out the most exciting news to a directorial listener, backing each statement with the shouted sentence, "Mind, I'm telling you this in the *strictest confidence*!") Partitions should be chosen as carefully as directors. There are a number of insulating boards which can cut off sound and which are used for telephone booths and other offices where quietude is essential. Infinite damage can be done to the efficiency of any organization by under-currents of sound; half-heard telephone conversations, bursts of raucous laughter; the drone of a manager dictating; the machine-gun fire of typewriters; the burr and rattle of an adding machine; and the bright young voices of typists at tea. All these things can be cut out with insulating board, by the use of surfaces which absorb sound, by the introduction of carpets with thick felt underlays.

The problem of carrying daylight across a floor that is divided by partitions is simplified by the use of such diffusing materials as kaleidoscopic glass and by keeping not only walls in pale colours but by having furniture in light tones. The clean, easily controlled artificial light provided by electric current has made it possible for the modern office to be a clean and comfortable place. Dust-proof electric light fittings are available, and cool globes of light can depend from the ceiling and shed an almost shadowless illumination. The adjustable desk lamp probably increases the output of work by its almost hypnotic power of concentrating the mind of its user on the job of work he or she happens to be doing.

The furniture of the office has inherited some of the dismal conventions of the Victorian era. For years metal furniture was made only in the most drab and depressing shades of khaki. It is difficult to know why such a colour was selected. It is repellent, cheerless, and carries with it a hideous suggestion of an institution: a workhouse, a prison, a lunatic asylum or (what practically amounts to the same thing) a war department. Metal furniture can be made in other shades, and cream and buff are cheerful and thoroughly practical as the hard-baked enamel surface can very easily be cleaned. In an office designed a few years ago a complete colour system for each department was worked out. Each department had its own coloured forms and the metal furniture and the woodwork in the offices allotted to that department were in the same shade as the forms, excepting only the accounts department where blue forms were used, and as solid masses of blue for cupboards and filing cabinets would have been rather cold and heavy, white furniture was used with a blue line on it. Tubular metal chairs are perfectly adapted for office use. They displace less space than the heavier wooden types, they look neat and are comfortable, and all is not chromium. The metal furniture provided by various manufacturers is simple in form, eminently fit for its job, and its plainness is commendable. Some people obviously feel unhappy about plain surfaces because some manufacturers can and do provide metal furniture, filing cabinets, lockers and so forth, grained to imitate mahogany. The disguise complex is persistent.

The equipment of the modern office is simple and compact; power, communication, heating and lighting can all be rendered simple. But there are office systems, systems of filing, systems for routine work, systems controlling the general circulation of activities, and unfortunately these systems are not always considered when accommodation is planned. It is the designer's responsibility in the first place to master in detail the system upon which an office or group of offices is going to be run, so that the whole space at his disposal is logically utilized.

The character of a firm's activities should be reflected as far as possible in the decorative treatment of its offices. It is not suggested that the reception-room of a firm of whiskey distillers should be fitted up as a bar with chairs in the form of bottles, although sillier things *are* being done in reception-rooms. But if a firm is manufacturing an article that can be used in decoration, then it is the worst kind of modesty to refrain from using that material. We have been into the offices of more than one firm which was making materials that could be used on walls and floors and we were shown samples of the materials, carefully taken out of a cabinet, when the walls and floors of every room in the place could have been a much more convincing and memorable exhibit of samples. There is still among a vast majority of English commercial concerns a reluctance to explain what they are selling. Sometimes this reluctance is based upon the assumption that everybody knows all about them. First impressions are important. Why do so many businesses appear to regard a visitor as an enemy? The reception-rooms and waiting-rooms of even the most enlightened firms are often outrageously furnished and decorated. Often you are thrust into a little cubby hole to await the pleasure of announcing to the controllers of the firm that you have got some business for them. It is related of an eighteenth-century gentleman that he kept a particularly repellent room in his house with hard wooden benches which he called the Jerusalem Chamber. Jews who used to dun him for money were shown in there to wait until they were too discouraged by the hard benches and the black outlook to wait any longer. Something of that fear of unwelcome visitors must be responsible for the character of hundreds of reception- and waiting-rooms. *Discourage the visitor: he may ask you for something!* is the archaic thought that perpetuates this incredible piece of commercial stupidity.

Often before you arrive at the reception-room you are incarcerated in a mahogany cavern which has been tricked out to look like a Sheraton wine cooler or something equally inappropriate. This is the lift. The individual firm that occupies a floor of a modern office building cannot control the design of the lift; but the architect of such a building can, and so often it seems to have escaped his control. An idiotic waste of money is apparent in many buildings. Once when a lift had broken down, the writer ascended to the sixth floor of a building by a staircase that had been designed to impress those who used it. Actually nobody used it; but money had been poured out on panelling it with Travatine marble and enriching its balustrade with costly metalwork. This unused, expensive staircase had been put in a six-floor building; but there was only one lift. True, that lift was very rich. It had an intricately carved frieze (fruit chiefly); but the building needed a battery of lifts, not a marble staircase.

In *Vitruvian Nights* Professor Goodhart-Rendel puts down some reflections about offices which suggest that there is no immediate prospect of that section of industrial design that is concerned with office equipment deriving much inspiration from contemporary architecture. "Most office-buildings are nothing more than workshops," he writes; "so in another sense are school-buildings, and the designation need not be greatly stretched to include public libraries. Architectural display is appropriate to exceptional buildings in all these classes, but the general run of them is better with no pretence beyond comely efficiency. In so far as the 'functionalist' style really functions, it could teach us to put all these workshop-buildings out of fashion's reach, to establish for them types that nothing but increase in engineering skill could be allowed to modify. This may be hoped for, but I fear that it cannot be expected. It is very difficult to make simple and convenient office-buildings expensive enough, either in first cost or cost of upkeep, to satisfy the importance of most big business firms, and I fancy that the present competition in columns and domes can be stopped only by the ridicule that public criticism is not yet ready to bestow upon it. In the main, with some honourable exceptions, the architects of big office-buildings will probably continue to bounce about from one elaborate style to another for some time to come."

Since the war, office accommodation in London and in some provincial cities has been rebuilt very largely. In London it has been rebuilt almost entirely without reference to twentieth-century architectural technique. Our great merchants and traders labour in an assortment of inappropriate exaggerations of Greek and Roman temples or imitations of pavilions of the 1925 Paris Exhibition. This may, as Professor Goodhart-Rendel suggests, satisfy their importance, but it is conceivable that such surroundings have a debilitating effect upon the financial mind. If this is so perhaps a contributory cause of the crisis of 1931 comes belatedly into the light. The opportunity of rehousing commerce appropriately has been largely missed, although it costs less to build logically with the wonderful array of materials that this age possesses than it does to hide the light of the twentieth century behind the old clothes of the eighteenth.

CHAPTER VII

COMMERCIAL ART

COMMERCIAL Art is concerned with the distribution of goods. Its chief manifestations are:

1. Advertising.
2. The display of goods.

1. Advertising for the purpose of discussing commercial art may be subdivided thus:
   (a) Press advertisements.
   (b) Posters, streamers, window bills, signs.
   (c) Literature: booklets, leaflets and folders.
2. Display may be subdivided thus:
   (a) Shop windows and showrooms.
   (b) Exhibitions.
   (c) Tins, cartons, packages, containers generally.

Those who design for advertising, design in two dimensions. Those who design for display, design in three dimensions. Although display is a form of advertising, for the sake of simplicity it will be regarded here as a separate subject.

The business of selling goods has, during the last half-century, developed a complexity for which there is no historical parallel. There is nothing comparable in any previous civilization of which we have records with the machinery of modern marketing with its

153

developing power of influencing public opinion by means of advertising, which operates in the press, through the post, on the hoarding, through the cinema film and from foreign wireless stations. The actual display of goods, the way they are set out to attract and inform the potential customer, has also developed a technique of its own that is wholly the property of the present age.

The shop used to be principally a warehouse that provided facilities for inspecting goods. In the rather low-browed little shops in the streets of a mediaeval city a conventional sign or symbol conveyed the character of the shop to the passer-by; but you had to go inside before you became adequately acquainted with the merchandise. Only at great fairs were goods openly displayed to attract the customer. Fairs, such as the Lynn Fair, which was one of the periodical European trading events in the Middle Ages, were the forerunners of the big trade exhibitions that have become such a commonplace feature of modern civilization.

The exhibition habit, which began seriously in this country with the Great Exhibition of 1851, has contributed many potent ideas to the art of displaying goods. Within Paxton's great glass exhibition hall in Hyde Park there was a sense of design, a real orderliness of lay-out and display. But in this field design degenerated. An exhibition soon came to mean a group of buildings that were in themselves admirable pieces of engineering with the architecture hung on afterwards: "applied" adornment on structural bones. In these palaces of stucco and steel there was a fortuitous concurrence of objects. Machinery would gleam here and there amid palm stands; beautifully made mechanisms would operate between trailing clouds of geranium and petunia depending from the roof girders in moss-lined baskets of galvanized wire. It was also the age of the glass case; great expanses of plate-glass were elevated upon polished mahogany stands, and they screened all manner of goods. The pre-war White City which combined in unhappy vulgarity the attributes of the "Fun Fair" and the commercial exhibition, was thought, when the Franco-British Exhibition opened, to touch the highest standard of display and lay-out. Unfortunately the glittering congestions of that show were to be accepted as standards of display and arrangement until 1924, when, largely owing to the late Sir Lawrence Weaver, designers were given an opportunity of producing a tidy, orderly and lucid display of goods and services at the British Empire Exhibition. A year before that Wembley show there had been at Gothenberg in Sweden a remarkable little exhibition which had been designed by an architect, and which demonstrated the unity and individuality an exhibition could have when it was the work of a competent designer.[1]

It was an exhibition of a new type and it drew people from England and all over Europe to Sweden. Everybody was interested, and professional critics were unusually kind, and the English press gave far more attention to the 1923 Gothenberg Exhibition

[1] See Chapter III, p. 94.

than it usually gives to that order of foreign affairs. The campaign to make the British public "exhibition-minded" was under way in 1923, for the British Empire Exhibition was to open the following year, and Gothenberg showed on a small scale how well an exhibition could be designed and how easily it could transcend the strident jumble-sale-cum-fair of the pre-war White City tradition. It illustrated how a core of controlled exhibits could impart a lucid orderliness to the whole show and how a disciplined lay-out could enhance the individual effectiveness of commercial exhibits.

At Gothenberg all manner of ideas for dramatizing the display of goods and services were assembled, and in the national exhibits the dramatic presentation of facts that are usually entombed in the tabular setting of Government publications gave vivid pictures of Swedish social and industrial improvements. The whole exhibition was an advertisement for Sweden, given architectural form. It created thousands of enthusiasts for Sweden and things Swedish; but it also created a new technique of exhibition design by revealing that national and commercial exhibitions are really advertisements in three dimensions, and that when competent architects are employed to design such three-dimensional advertisements they are as different from the old, muddled, crowded pre-war exhibitions as a skilfully type-set and vigorously illustrated modern advertisement is from the hotch-potch of nondescript type-faces that shouted each other down in the majority of pre-war advertisements.

Whenever trained designers are given opportunities

by commerce in exhibition work, lucidity and order appear which make the exhibition memorable, and which increase the chances of attracting public interest in the things which are shown.

There is an annual event of importance to the public, wholly commercial in character, but in recent years, increasingly intelligent and attractive in lay-out. It is the "Ideal Home Exhibition," organized by the *Daily Mail* every spring at Olympia in London. Since architects and designers have taken a hand in the lay-out of the Main Hall at Olympia, the whole character of the exhibition has changed. Perhaps it has proved to innumerable firms who display their goods there that design pays. It is a difficult thing to prove. But presumably it is being proved, for in exhibition work increasing numbers of commercial organizations are becoming convinced that it pays to employ a designer to arrange the display of their goods. This used to be considered the job of the display man or the window dresser. The actual setting out of goods in a window, the arrangement of display sets, the piling up of packaged goods, tins and cartons, is a highly skilled job which has to be carried out by a display expert, but it is not the job of the display expert to solve all the problems of the economical use of a site in an exhibition. The imposition of standard schemes for the form and colour of stands in exhibitions naturally simplifies the problem of display.

Up and down the country during the year many trade exhibitions are held as well as those to which the general public are admitted. In London alone there are over thirty exhibitions in the course of a year. There are tobacco exhibitions; furniture exhibitions; (in London and Manchester); housing and health exhibitions, dairy shows, brewers' exhibitions, radio shows, confectionery, grocery, drapery and hairdressing exhibitions, to name only a few. Most of the principal shows are held in London, and they are nearly all duplicated or triplicated in the provinces.

Trade by trade, commerce is waking up to the fact that well-designed stands are easier to run, more attractive to the customer, more profitable to the stand-holder. Any architect who designs either an exhibition lay-out or an exhibition stand is designing an advertisement for something. Inevitably he must study the effect his exhibition stand or stands will have upon the potential customers for the products or services that are being sold from that stand. In collaboration with his client he will probably study that problem with the same assiduity as the people whose business it is to prepare advertisements study their problems of appeal and presentation, but unlike them the architect is not limited to two dimensions. Matters now engage his attention which twenty years ago were seldom his concern: the design of lettering and type-faces, for example. These are the materials of the typographer; but the architect who is designing advertisements "in the round" must use them. The display of goods in a shop window is the business of the professional window dresser; but the architect has to study that·branch of design also in exhibition work.

In ten years the amount of advertising designed by

architects at exhibitions has been considerable. The old go-as-you-please-shove-it-together-anyhow individualism has been banished from the big, well-organized shows.

In the designing of stands architects have concentrated on bringing order out of chaos. They have come into contact with the pride that nearly all manufacturers have in their products, a pride that fills them with a simple and often quite unjustified faith that the public knows all about those products and wants to hear as much about them as possible, and wants to see as much of them as their eyes can take in, and to study where and how they are made. This sort of pride makes manufacturers overcrowd their notepaper with lists of the things they make crawling up the left-hand margin, with bad drawings or worse photographs of their factories over or under the heading; and it controls their attitude towards any public appearance of their wares. Usually the architect who designs a stand for the display of those wares is exhausted after a stiff battle with his client on the matter of overcrowding, and if he attains simplicity and reticence, and succeeds in preventing the use on an enlarged scale of the name block that was designed for the firm in 1840 or thereabouts, he has done pretty well.

But the design of an exhibition stand is more than giving architectural form to a display. It is more than a compact and economical solution of the problem of showing materials. It has to be more than an accommodating statement of fact. It has to dramatize

159

the facts about the goods or materials, by repetition or imaginative interruption. Repetition, especially when it emphasizes some particular quality of the goods displayed, can be made most powerfully dramatic.

Stand design is not tied to any particular materials. It is as free from constructional limitations as stage scenery, and to regard an exhibition stand that has to attract a bored and capricious public as anything but stage scenery is to impose a severe check upon the inventiveness of a designer. New and stimulating combinations of glass and aluminium and steel and neon lighting and photographic enlargements, models, miniature cinemas, graphs, charts, pictorial statistics, can allow the most adventurous stand to hint at the enormous powers of attraction commerce can gain by employing designers who comprehend the possibilities of modern materials and mechanical devices and modern illumination.

In exhibitions a new form of collaboration between design (represented by architects) and industry could be and is being developed. It affords abundant opportunities for architects to inject into this branch of advertising fresh and vivid qualities, and the distinction attained by "architect-built" stands should impress everyone with goods or services to sell to the public, or, if they are manufacturers who operate through retail channels, to their own distributors. It may take some years of education before it is realized by commerce and industry that when you have paid some hundreds of pounds for space at an exhibition it is worth spending a little more money to see that it is filled intelligently. It took a long time to convince most advertisers that it was stupid to agree to spend eight or nine hundred pounds on a large space in a great newspaper and to kick at spending more than a guinea or two on the illustration that was to appear in it. It will pay exhibitors at trade and public exhibitions to employ a good architect; and if they are reluctant to deliver themselves into the care of one expert, then an easy and established method of selecting their architect is available. Let the manufacturer or retailer who is exhibiting make up his mind what he wants his stand to do, and then let him state his requirements as a problem, and arrange a competition for designs. It is well worth spending a hundred or a hundred and fifty pounds in prizes to secure some good architectural brain for the solution of a problem of display.

Advertising is a mechanical amplification of the old cry of "Buy! buy! buy!" The printing press has become its chief vehicle since a huge reading public came into existence in the last half of the nineteenth century. It commands the services of artists, typographers, writers, film and radio specialists, and scientific experts who explore with a laboratory technique the potentialities of different markets, so that the right objectives can be identified and the appeal of advertising psychologically adjusted to the character of those objectives.

Advertising to most people means posters. Posters are the most commonly memorable manifestations of advertising, but actually they represent a relatively small part of the business of calling attention to

161

various goods and services. Advertisements which appear in the press are responsible for a far greater proportion of the output of commercial artists, and it is the press advertisement which has produced what may be described as the advertisement designer. An advertisement designer is concerned with the presentation of an advertisement, that is to say the working out of the relationships between masses of type and illustration and the illumination of the whole theme with an advertising idea. The advertisement designer works in collaboration with the copywriter, and an advertising idea is often the joint product of those two widely differing types of creative mind.

Making an advertisement is not confined to the task of using words skilfully and economically and then associating them with appropriate illustrations. A knowledge of the use of words is essential, and the possession of this knowledge and the ability to apply it is not common. The late Arnold Bennett, in his book *How to Become an Author*, wrote down this piece of wisdom: "The first sign of unintelligent writing, the first cause of tediousness, is the presence of ready-made trite phrases."

People who write books and articles do not always remember that; but the author or journalist who inflicts ready-made phrases and unskilful writing upon his readers is certain of some kind of audience; people have paid for his book or the paper or magazine in which his article appears, and they may not only read what he has written but they may be quite uncritical about its style or the occasional patches of tediousness that make his pages dull and long-winded. His readers are in a receptive mood; they have bought what they hope is going to be entertainment and interest or news or instruction.

The writer of an advertisement has got to get readers under very different circumstances. An advertisement dare not be tedious; otherwise it is wasted. An advertisement dare not be unintelligent or in any way obscure; otherwise it is a disadvertisement for the firm that issues it. An advertisement is not a good advertisement because it pleases the people who issue it; because it seems a dignified announcement; because it looks important and gives the name of the firm in large type and assures everybody who cares to read that Blank, Dash & Company, Ltd., have for the past half-century been renowned for the careful attention they give to all orders, for the high standard of their service and the unfailing quality of their goods. Such statements, truthful and dignified though they may be, are only capable of arousing one emotion—boredom.

An advertisement has got to fight for its readers, and the battle is not a battle of words alone. An advertisement should be the dramatization of an idea. Directly the word "idea" is mentioned in connection with advertising, intricately clever stunts are apt to be suggested. But a good advertising idea is a thread upon which whole campaigns of advertising appeal may be strung without the thread wearing thin. The idea that vitalizes any advertising theme must be of the kind that is capable of simple expression. The

163

most brilliant example of a graphic idea that has appeared during recent years is the Shell advertising; the selected selling point of the product being dramatized unforgettably by the series of figures registering astonishment at the speed of Shell-using motorists. The double-headed figures, indicating the quick turn of the head from right to left, the exclamation: "Crikey! That's Shell, that was!" drove home the idea of breathless speed with dramatic simplicity.

Advertising has, since the war, developed a variety of techniques which were unknown in pre-war days and which have enabled many artists to work very closely with the printing industry, and by studying its processes and the various methods of reproducing drawings and photographs, to achieve entirely new forms of illustration.

Before the war a printing revival had begun and it gained its momentum because advertising provided the patronage and enabled innumerable experiments to be made. A number of forcible personalities were connected with this movement, men like Sir Herbert Morgan, Joseph Thorp (who has written by far the most illuminating book that has ever appeared on the subject, namely *Printing for Business*), R. P. Gossop and many other artists who had taken the trouble to understand the printing press, and who made it work for them as it had never worked before. When in the early years of this century such men set out to improve standards of design in printing and in advertising, they discovered that type-faces in the nineteenth century had suffered from the general uglification that had debauched the form of nearly every physical object. Letters of the alphabet had acquired a squalid corpulence, a needless brutality or else a spidery complexity. In the latter part of the nineteenth century William Morris had it is true endeavoured to improve standards of type design; but to him the printing of a book was an occasion for glorious decoration rather than for the lucid austerity of legible type. In the seventeenth and eighteenth centuries type-faces had been based on the classical letter forms that were incised upon the Trajan Column at Rome, and, as it is stated in that admirable book on typography, prepared, printed and published by the *Pelican Press*: "What makes type more or less legible, and so, in that measure, beautiful, is the degree in which it conforms in its proportions to those great artificers who made the Roman inscriptions and inspired the letter-designing geniuses of all succeeding centuries; the artificers who made the code we daily use and made it technically perfect."

Consider this word which is set in Caslon Old Face:

INDUSTRIAL

The letters used in that word are all based upon the original Roman letter forms, and these forms are capable of considerable variation without any diminution of legibility. For example when the word is set in Open Titling:

INDUSTRIAL

it acquires a certain luminosity, a radiance and a gaiety which leave unharmed the individual character of any of the letters. They still remain clear, rapid signals of sound.

When a little more freedom is taken with the letter forms, providing the proportions are retained, then legibility does not depart. Here is a word set in Imprint:

INDUSTRIAL

If extra boldness is desired again the right proportions should be retained. The word is now set in Heavy Old Face:

INDUSTRIAL

Here in Garamond Bold:

INDUSTRIAL

Or, if something lighter is needed, in Garamond:

INDUSTRIAL

But in the nineteenth century and in the early part of the present century attempts were made to create types which were "different" or decorative by monkey-

ing with the proportions of the letters with the sort of
unhappy results visible here:

INDUSTRIAL

Observe the top-heavy and unhappy letter R.
Worse things were done to this alphabet than the

word printed above reveals; for instance: G C
K F E. The printing revival did re-establish
legibility, it re-established respect for the proper proportions of the alphabet so that when after the war
the printing revival entered upon another phase and
was influenced by the modern movement, and type
designers wished to take liberties with the old letter
forms, they were far better equipped to do so than the
Victorian type designers who were merely concerned
with spurious novelty. The post-war Continental type
designers who were chiefly active in Germany were
concerned with making dramatic and vigorous signals.
They produced the Erbar family, which achieved a
rapid popularity. When the character of Erbar is
analysed the reason for its popularity is easy to understand. It was conceived by German type founders in
a mood of vigorous disapproval of the old complex
Gothic text that had undermined the national eye-
sight of Germany for years. The traditional Gothic
text was favoured by that romantic European nuisance,
Kaiser William II. Shadowy illegible letters went with

waxed moustaches, gilded helmets on which eagles perched with precarious violence, swords which rattled with cinemaesque exuberance, and fists, mailed, one, militarists, for the use of. The stuffy, romantic, traditional taste of pre-war Germany belonged to a culture that was as dead as the dodo and as dangerous as a mad dog. (Maybe it will be revived now that mass hydrophobia has again been organized beyond the Rhine.) German designers were sick of everything that represented Hohenzollern taste. Infuriated by the illegibility of the traditional types, they created a mechanistic alphabet in which each letter was a clean, sharp symbol. To German eyes the new sans serif types both for displayed lines and for text were wonderfully legible; they represented a change as great as the change from a dark room to one flooded with sunlight. The result is best illustrated by Erbar Light:

INDUSTRIAL

Very different is the type designed by Mr. Eric Gill. Mr. Gill has created his alphabet with English eyes and with the skill of a sculptor and master mason who has for many years cut the most beautiful lettering in stone, based upon the Trajan Column alphabet. His sans serif type is not just the Roman alphabet with the serifs trimmed off; it is the individual solution of each letter problem, based on giving the simplest expression by a familiar shape, as opposed to the German method of intentionally destroying the soft familiar features of

letters and getting back to their hard bones. This is
set in Gill Sans:

INDUSTRIAL

Compare the R in Gill with the R in Erbar and the
G in Gill with the G in Erbar.

For the purposes of advertisement display many
bold decorative types have been resurrected from the
eighteenth or early nineteenth century or invented to
tickle the fancy of advertisement designers. Some of
these types are so urgently decorative that their use
except for fairly short display lines could not attempt
to attain legibility. Broadway is an example:

INDUSTRIAL

Mere stunt-mongering in letter forms is on the wane,
except in a few advertisements. But a few years ago
every national newspaper was alive with crawling
blackamoores, innumerable mis-shaped negroid letters
jazzing their way through advertisements, and they
were so self-consciously new, so vigorously novel, that
they frequently conveyed a shock without conveying
any sense.

But both Erbar and Gill demonstrate the basic
soundness of their letter forms, when they are varied

in weight and character. For instance this is Gill Sans Bold:

INDUSTRIAL

This is Gill Sans Extra Bold:

INDUSTRIAL

This is Gill Sans Extra Light:

INDUSTRIAL

This is Gill Sans Shadow:

INDUSTRIAL

It has been necessary to show these examples of type variation in order to suggest the scope that designers have had in only one of the materials connected with advertising. We have dealt only with capital letters and not with lower-case letters, only with types in display sizes and not with text types. This book is printed in a type called Baskerville which was invented by an Englishman of that name in the eighteenth century, but it might have been printed as the next sentence is printed in Gloucester Old Style:

A lot of this sort of type would be irritating to read, and yet people were irritated in this way by printers in the nineteenth century. Again, we might be irritated in another way, as the next sentence demonstrates.

This sentence is set in Clarendon, and it is not very kind to the eyes.

Typography is only one of the sections of commercial art which has been able to attract the interest of designers. It is beyond the range of this brief sketch to examine in detail all the varied manifestations of commercial art. Since the war press advertising has changed in character. It has benefited by the direction of many able designers, who have abandoned the convention that an illustrated advertisement consists of a picture with a headline, and some copy that is in the nature of a caption to the picture, the whole inert structure resting upon the uninspired base of some unalterable name-block. It has been said that this static type of lay-out was analogous, in proportion and the disposition of its masses, to the Roman orders of architecture. Now the Roman Doric, Ionic and Corinthian orders have three broad divisions: the plinth or pedestal, on which stands the column supporting the entablature (the latter subdivided into architrave, frieze and cornice). These three horizontal divisions were held to apply to the perfect lay-out. The name-block and standing matter, address lines and so forth, were analogous to the plinth; the type mass of copy to the column space; and the headline and subheads to the entablature. Such analogies suggest a somewhat costive reverence for fixed proportion, preserved at the expense of understanding the principles of architectural composition, which make the proportions of the classic orders articulate by resolving them into masses whereby their formality acquires a collective significance.

The static, beautifully balanced and symmetrical advertisement lay-out is tremendously gratifying to the man who makes it. Actually he is being as dully uninventive as a pre-war architect, who lacquered some architectural style based on the classic orders over the front of his building. The neat and static advertisement lay-out can be, and often is, beautifully legible and completely visible. It is often acceptable to the taste of the connoisseur of typography in its well-chosen association of headline and text. But the modern lay-out aims at interruption without irritation, and cuts with the sharpness of a sword through all the rigid trappings of tradition; it becomes a flashing statement in which type and illustration work together, not in any set relationships, but in some graphic arrangement, unforgettably lucid.

It is to such designers as Ashley Havinden and E. McKnight Kauffer that we owe the development of this particular technique in this country. For this new movement in advertisement design there are materials in stimulating abundance. The mechanistic clarity of the Erbar type-face has meant almost as much to the designer of modern advertisements as reinforced concrete has meant to the designer of modern buildings. That accommodating and vigorous Erbar family with its various weights, its condensed, lined and italic variants, enable every inflection of tone to be conveyed in displayed lines of capitals. Masses of lower-case Erbar used for panels of text tend to become illegible, and the association of Erbar display lines with the text in Plantin Old Style, Old Style Antique or Typewriter type, has produced many memorable advertisements. But all is not Erbar. A criticism, which has now become old-fashioned, that "modernism" in advertisement design depended upon the use of German typefaces, has been gradually deflated, as year after year it has become apparent to anybody with any pretentions to a critical faculty that the modern movement in advertisement design depends not upon the particular materials that are employed, but upon the manner in which the whole problem of design is approached. This is where the advertisement designer differs fundamentally from the modern architect. If the architect was deprived of his modern materials, of structural steelwork, and concrete and reinforced glass and rubber, and electric power to heat his slender transparent towers and to operate his batteries of lifts, there would be a perceptible contraction in the range and boldness of his work, although he might still attempt to do new and exciting things.

The absence of all the flexible materials that are used in this second third of the twentieth century would hamper his powers of expression, would curb and change the whole scale of his work. But the advertisement designer, even if he were restricted to the use of types that were available in pre-war days, would still be able to express the vigour of the modern movement in design.

At present a far greater number of competent designers find employment in connection with distribution than with production. Advertising has provided abundant and encouraging patronage for design. Commercial art is healthy and vigorous and has attracted some first-class talent. When industrial art attains the same standards of health, and encourages its practitioners to develop the questing, experimental outlook upon their work that distinguishes the men who do the best work in advertisement design, we shall live in a new age, comparable in achievements, in orderliness and universal comeliness to that golden age of design, the eighteenth century—but far more exciting.

CHAPTER VIII

THE FUTURE OF THE DESIGNER

> "'Painting, sculpture, all furnishing and decoration, are the escaped subsidiaries of architecture, and may return very largely to their old dependence."
>
> *The Work, Wealth and Happiness of Mankind,*
> by H. G. WELLS

THE architect is almost the only professional type of man in the community who is trained to think lucidly about design. He is trained to employ a perplexing diversity of materials for carrying out his work. His ability to plan may give him some exceptional opportunities for affecting the life of the community in the future, not only in the obvious directions of building and town-planning, but by forming the character of industrial products, and by influencing the methods chosen for distributing such products to the general public.

In the past the patronage for design has come principally from three sources. Either from an established institution, such as the Church, or from the head of the State, or from wealthy and powerful individuals. The last source of patronage has, since the Middle Ages, provided most of the work for English architects. When the wealthy individual clients were educated, and in consequence took a lively and informed interest in design, great periods of architecture became possible.

175

From the time when the true rhythms of the Renaissance were first revealed by Inigo Jones, until the reign of William IV., England enjoyed a period of order and urbane beauty that we look back upon with admiration; and sometimes, after inspecting the discords of our own age, we may look back to those Stuart and Georgian lucidities with a feeling of despair. Many of our present troubles in architecture and design we owe to gross or timid patronage. It is instructive to examine the problem from the designer's point of view.

Anybody who is studying architecture and who hopes eventually to practice, will soon be confronted with the peculiar conditions of contemporary patronage. An architect's training gives him an authoritative background and a feeling of independence about materials; independence in the sense that he is not browbeaten by materials, nor obsessed by their limitations, although he understands and respects their limitations.

The command of architects over materials has been mentioned here because when they begin to practice they will find themselves in a world which resolutely refuses to acknowledge that there has been a structural revolution in the last half-century. Knowing what they could do with steel and glass and concrete; envisaging the splendours of form they could command with the materials of this age, they will be nearly always compelled by their patrons, and if they are building in London they will often be compelled by restrictions upon elevations, to deal in archaic forms, and to use costly, needlessly heavy and thick materials for the clothing of their steelwork. This will change. It is already changing. There are tentative experiments which suggest that there may, in due time, be big changes in the spirit of patronage and perhaps an intelligent remodelling of building regulations. But those who would take comfort from one of the few buildings in London, the *Daily Express* building in Fleet Street, which suggests that patronage is getting bolder, and is actively encouraging experiments, would do well to remember that the sort of patronage that made the *Daily Express* building possible is exceptional. It is conceivable, as education moves so slowly, that the coming generation of architects may not have their opportunity for twentieth-century design until London, Manchester, Birmingham, Liverpool, and a few other cities and residential areas have been levelled by the air raids of another war. The opportunity for architectural design after this event may be limited by a shortage of clients, or even a complete absence of civilization. Assuming that the present system is going to lurch along without such nonsensical interludes of destruction, the architect's opportunity for creative design may be classified roughly under four headings:

1. The wealthy individual client who wants an individual house, large or small, country or suburban.
2. The public utility company or municipal corporation which is concerned with the erection of public buildings and housing schemes.

3. The financial development company which is concerned with the exploitation of residential areas and sites, the erection of housing schemes, individual houses and blocks of flats.

4. Industry, which is concerned with the production of goods, of which many types were once controlled in their design by architects, and with the commercial distribution of goods.

The wealthy client is going to be scarce in the future. He is scarce now. He will ask for the adaptation and the modernization of existing buildings instead of initiating fresh work. With more and more country houses going out of commission and with the gradual redistribution of income which is happening without socialism, communism or the application of any predatory political faith, there will be far more camping-out done in parts of old houses, far more using up of things that exist, and an acceleration of the movement of people into ready-made houses and flats. This will cause financial development companies to expand in importance as patrons. Whether the ideas of such bodies are eventually controlled by the architectural profession depends very largely on whether architects can convince accountants and other money-minded people that there are no risks attached to good design and that it could even pay. The onus of justifying his work financially is also upon the architect who works for a public utility company or corporation or for municipal authorities.

Although there will be a·great deal of building done

during the next ten years, and, one hopes, an enormous amount of poisonous and disgusting rubbish cleared away, an equally great opportunity awaits architects in the activities of industry. Not merely in the building of factories or premises for distributing trades, nor in the extensive reconditioning of workshops everywhere. Electrical development will be changing industrial technique, and factories can be made cleaner and more orderly. Apart from that, in the vast business of manufacturing commodities a few men in control are beginning to recognize that industry lacks an important branch of technical advice. Inevitably the architect must come into industry as the designer, as the missing technician.

Industrial production as we have seen is ably conducted by what may be called process technicians, engineers and chemists. Creative industry in the near future may offer to the architect the opportunity of designing, not buildings, but goods. The architect could become recognized as *the* expert on design in the community, and as such he could regain the control over the form of everything that he possessed in the eighteenth century. Already some enlightened manufacturers of certain types of building materials have appreciated the importance of employing architects in a consultancy capacity on all matters concerned with design. The practice of co-operation between designers and manufacturers is growing slowly, and *the architect is the only designer whose training is such that the manufacturer can be certain of getting the right sort of expert advice on design.* It is the architect's ability to understand the

179

limitations of materials and to grasp their character that commends him to the manufacturer.

For a young architect there could be few studies quite so profitable, after he had completed his architectural training, as the nature of different forms of industrial production. The architect, if he is to control design once more, should know how furniture is made in a modern factory; should know how pottery is made; how carpets are produced; how glass is made; how metalwork is fashioned. He needn't spend years of his life in factories; but familiarity with even the superficial aspects of industrial technique, will enhance his ability as a designer, and what is equally important his *authority* as a designer.

Already architects are beginning to be called upon to design furniture, radio and gramophone cabinets, display fittings for shops, all the apparatus used in the retail distribution of goods. Not only will the big industrial firms require consultants about design, but the enlightened retailer will need the same sort of guidance.

Here is an example. Electrical development has already been mentioned in connection with improving industrial technique. Unfortunately the provision of an abundant supply of electricity is no guarantee that it will be used. The producers of electricity are in exactly the same position as a manufacturer who makes goods. Those goods have got to be sold. The manufacturer distributes them either direct or through a retailer, to the ultimate consumer. Electricity is distributed to the consumer through local supply companies.

Now local supply companies, up and down the country, have got showrooms where various types of apparatus are displayed which can be hired or purchased by the domestic consumer of electricity. There is much work to be done by architects in improving the design of such apparatus, as well as in improving the design of the surroundings in which it is displayed. A number of electrical supply companies are now modernizing their showrooms. Two conspicuously interesting examples of this are in London. The Kensington Electric Supply Company, whose showrooms have been designed by Mr. Raymond M'Grath. Another and more recent example is the Westminster Electrical Supply Corporation, whose showrooms in Victoria Street, London, have been designed by Mr. E. Maxwell Fry.

A much more difficult problem is involved in the redesigning of electrical apparatus. Electrical fittings have only just begun to be well designed in this country. Only a few manufacturers are interested in making good shapes. Even now adaptations are made of fittings originally designed for candles. Electrical apparatus, as an earlier chapter demonstrated, has seldom been allowed to behave properly. The amount of junk which is still being manufactured is depressing, such as heaters which attempt to imitate logs and burning coals, and cookers elevated on white enamel cabriole legs. The firm that is doing the best work in electrical equipment and lighting to-day, has on its board a trained designer. We can, perhaps, look forward to a future when architects will serve on the boards of companies which are concerned with manufacturing apparatus of every description, and when the architect will be nationally recognized as the proper consultant on all matters pertaining to design.

Now we come to the problem that business firms have. Let us say they have become convinced that good design pays. Where can they get good design? The answer is: From the young men in the architectural profession.

Now who are the people who are going to control or at least influence this branch of commercial and industrial patronage? There is a growing desire to get better work designed not only in display on the distribution side, but in production on the manufacturing side. There is probably a bigger demand for good design than we can possibly assess. But the great difficulty of industrial patronage is this.

The man of business, who has been laughed at and rather despised, knows his job, and generally speaking knows how to use specialists. Even when he has recognized that the designer is a specialist, he is not capable of discriminating between a good design and a bad design. The business man's problem of selection in design is a difficult one. If he is educated, and his eyes are trained to distinguish the difference between good proportion and bad proportion, then he is a very rare bird indeed. Generally he does not realize that there are many shades of excellence in design, dulling down through mediocre effort to frank atrocity. It is often difficult for him to understand that he is not employing a mere technician when he is employing a designer. Still in this matter we must hope for some effect from the barrage of propaganda which has been hammering at commercial and industrial administrators now for the best part of ten years; and which has culminated in the 1935 Royal Academy Exhibition of Industrial Art. Everywhere the man of business is being made to realize that somehow or other he has missed the boat so far as design is concerned.

The Chairman of a great manufacturing firm in the Midlands once said to the writer: "We are perfectly prepared to use designers if they are prepared to be practical. I don't mean by being practical that we want them to go to an office at 9 o'clock in the morning and behave like clerks. Many men in business to-day take that point of view; but what I mean is that we want the designer to study our problems of production so that he does not set us impossible tasks." He mentioned two unfortunate experiments he had made of employing designers and of one successful one. The successful experiment was the last one he had made, for he had not been rendered impatient or sceptical by two failures. The third designer had been successful because he had taken the trouble thoroughly to understand the special processes which were involved in producing his designs, and the result was an alliance between the creative man and the willing machine which was extremely satisfactory to both the man of business and the designer. Incidentally the successful designer was an architect. The other people had been studying what some art schools call "art." It is perhaps unjust to say that the art schools have failed, but surely the only school that can produce designers who will mean anything to this machine age, is the school that trains them to handle materials and strengthens their perceptions and expands their invention by making them study how, in the past, materials have been handled and how, in the present, contemporary materials may be mastered. Most architectural schools do this, and some art schools. But that Academy of Design, mentioned in the Introduction, which could reinforce the work of the schools by stimulating the education not only of industry but of the public, is still a dream.

Meanwhile industry is slowly becoming educated. It is better able to appreciate its weak points to-day than it was five or six years ago. When most of its products are well designed, we shall discover that we are in the beginning of a new Renaissance. But that can only come about by a real partnership between design and industry. It is only the architect, the trained designer, who can make that partnership valid. It is only the public, who can make it permanent by being more critical about the things they buy, thereby sustaining and enlarging the demand for good design, so in time the old lost standards of common art may be resurrected in English Industrial Art.

APPENDIX

A LIST OF BOOKS ON DESIGN

THIS is not a bibliography. It is a short list of books, most of them extremely readable, concerned with various aspects of industrial art and machine-craft, with architectural design, with planning, and with patronage.

The Appreciation of Architecture, by Robert Byron. (Wishart & Co.)

Architectural Style, by A. Trystan Edwards. (Faber.)

Architecture, by Christian Barman. (Benn's Sixpenny Library.)

The Architecture of Humanism, by Geoffrey Scott. (Constable & Co.)

Art, by Clive Bell. (Chatto & Windus.)

Art and a Changing Civilization, by Eric Gill. (John Lane, Twentieth Century Library.)

Art and Counterfeit, by Margaret Bulley. (Methuen.)

Artifex, or The Future of Craftsmanship, by John Gloag. (Kegan Paul: To-day and To-morrow Series.)

Art Now, by Herbert Read. (Cassell.)

Balbus, or The Future of Architecture, by Christian Barman. (Kegan Paul: To-day and To-morrow Series.)

Beauty Looks After Herself, by Eric Gill. (Sheed & Ward.)

Craftsmanship and Science, by Prof. Sir William Bragg. (Presidential Address to the British Association, 1928. Watts & Co.)

Design in Everyday Life and Things, The Year Book of the Design and Industries Association, 1926–27, edited by John Gloag. (Benn.)

Design in Modern Life, edited by John Gloag with contributions by Robert Atkinson, F.R.I.B.A., Elizabeth Denby, E. Maxwell Fry, B.Arch., A.R.I.B.A., James Laver, Frank Pick (President

INDUSTRIAL ART EXPLAINED

Design and Industries Association), A. B. Read, A.R.C.A., and Gordon Russell. (George Allen & Unwin.)

Design in Modern Printing, by Joseph Thorp. (Benn.)

The Economic Laws of Art Production, by Sir Hubert Llewellyn Smith. (Oxford University Press.)

English Furniture, by John Gloag. (Black's English Art Library.)

Everyday Architecture, by Manning Robertson. (Fisher Unwin.)

Form and Colour, by Lisle March Phillipps. (Duckworth.)

Form in Civilization, by W. R. Lethaby. (Oxford University Press.)

Furniture and Furnishing, by John C. Rogers. (Oxford University Press.)

Ghastly Good Taste, by John Betjeman. (Chapman & Hall, Ltd.)

Good and Bad Manners in Architecture, by A. Trystan Edwards. (Philip Allan.)

The Honeywood File, by H. B. Creswell. (The Architectural Press.)

Machine Art, by Philip Johnson. (George Allen & Unwin.)

The Meaning of Art, by Herbert Read. (Faber.)

Men and Buildings, by John Gloag. (Country Life, Ltd.)

The Mistress Art, by Sir Reginald Blomfield. (Arnold.)

Modern English Architecture, by Charles Marriott. (Chapman & Hall.)

Modernismus, by Sir Reginald Blomfield. (Macmillan.)

The Origin of the Sense of Beauty, by Felix Clay (1908). (Smith, Elder & Co.)

The Pleasures of Architecture, by C. and A. Williams-Ellis. (Cape: Life and Letters Series.)

The Principles of Architectural Composition, by Howard Robertson. (Architectural Press.)

Printing for Business, by Joseph Thorp. (W. H. Smith & Son.)

The Revolutions of Civilization, by Sir W. M. Flinders Petrie. (Harper & Brothers.)

Room and Book, by Paul Nash. (Soncino Press.)

Time, Taste and Furniture, by John Gloag. (Richards Press.)

APPENDIX

Vision and Design, by Roger Fry. (Chatto & Windus: Phoenix Library.)

Vitruvian Nights, by Prof. H. S. Goodhart-Rendel. (Methuen.)

The Works of Man, by Lisle March Phillipps. (Duckworth.)

JOURNALS AND ANNUALS

The Architectural Review. Monthly. 2s. 6d.

Commercial Art and Industry. Monthly. 2s.

Design for To-day. Monthly. 1s. (The Official Organ of the Design and Industries Association.)

Die Form (German). Monthly. RM. 1.

Form (Swedish). Monthly. 10 kr. a year.

The Studio. Monthly. 2s.

The Studio Year Book.

INDEX

INDEX

P L A T E I.—Industrial architecture in the eighteenth century. Two mills in the Stroud Valley. Above: Hope's Mill at Brimscombe; below: Bentley's piano factory at Woodchester. These buildings were conceived in a strong existing tradition of architecture. (Photographs reproduced by courtesy of Thomas Falconer, F.R.I.B.A.)

P L A T E I I.—*Industrial architecture determined by the needs of a particular manufacturing process. The Old Casting Hall of the British Cast Plate Glass Company who established their works at Ravenhead, St. Helens, in 1773. (Reproduced by courtesy of Pilkington Brothers, Limited.)*

P L A T E I I I.—*The new industrial architecture of steel, concrete and glass. Part of the factory at Nottingham designed for Messrs. Boots by Sir Owen Williams, K.B.E. Like the Stroud Valley mills on Plate I. and the Casting Hall opposite, this building shows contemporary materials being used to the best advantage.*

P L A T E I V.—Machine architecture. The exhaust fans in one of the ventilating stations of the Mersey Tunnel. (Architect: Herbert Rowse, F.R.I.B.A.)

P L A T E   V.—*An example of machine design. A glass insulator made by Pilkington Brothers, Limited, St. Helens.   Shapes such as these, determined by a mechanical need, frequently inspire and stimulate the ideas of designers, but they remain examples of machine design. Forms evolved in the process of solving an engineering problem naturally afford relief to eyes wearied by industrial productions in which function is disguised; but such forms do not contribute anything to the advancement of industrial design.*

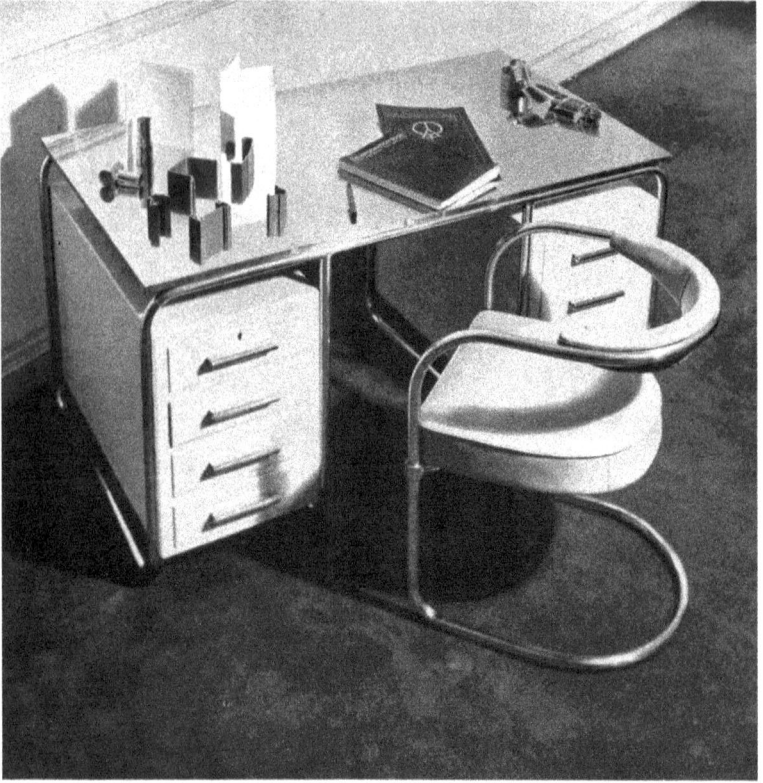

PLATE VI.—The influence of new materials and processes upon industrial design is here illustrated by a desk and chair with arms of copper tubing. The top of the desk is of copper Plymax, which is plywood faced with a sheet of copper, and the nests of drawers which are clasped by the tubular copper frame are cellulosed a pale cream colour.

PLATE VII.—*Another example of new materials influencing the design of furniture and of interior architecture. The counter and screen in this bar are in pearl-grey Vitrolite, thin lines being acid-embossed and coloured in cerise to match the carpet. The door consists of a single sheet of pearl-grey Vitrolite. The counter top has a stainless steel edging, and the plate-glass screen on the counter has a stippled ground with three lines left clear. The table top is in primrose Vitrolite with a design in shaded sand-blast on the top. The chairs and the table have frames of chromium-plated tubular steel. (Reproduced by courtesy of the British Vitrolite Company, Limited. Architect: Kenneth Cheesman.)*

P L A T E V I I I.—*Above: An Ultralux electric-light fitting designed by A. B. Read, A.R.C.A. The flashed opal glass globe has a thread in the upper part which screws into the chromium-plated metal frame and makes a dust-proof joint; a simple form of light fitting which precludes glare. A good example of industrial design when it comes under the control of a competent designer. (Reproduced by courtesy of Troughton & Young, Ltd.)*

P L A T E I X.—*To the right: An example of new materials enabling an architect to create an external effect in which artificial light is in flexible alliance with metal and rubber and plywood. Here Neon lighting is used for the lettering on the fascia and for the lines that lead down from it and run above the small display window. (Showrooms for an electric supply company, by E. Maxwell Fry, B.Arch., A.R.I.B.A.)*

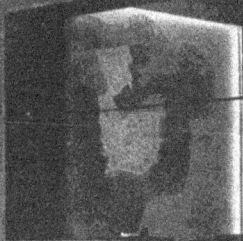

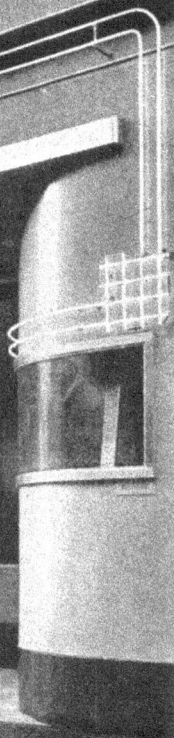

P L A T E X.—*Two examples of decorative glass. Above: the " Cascade " pattern, designed by R. A. Duncan, A.R.I.B.A.; below: " Coptic," designed by Paul Nash. (Both designs manufactured by Chance Brothers & Co., Ltd.)*

P L A T E X I.—*A wood block print on linen in six colours designed by Paul Nash for Cepea Fabrics. (Calico Printers' Association.)*

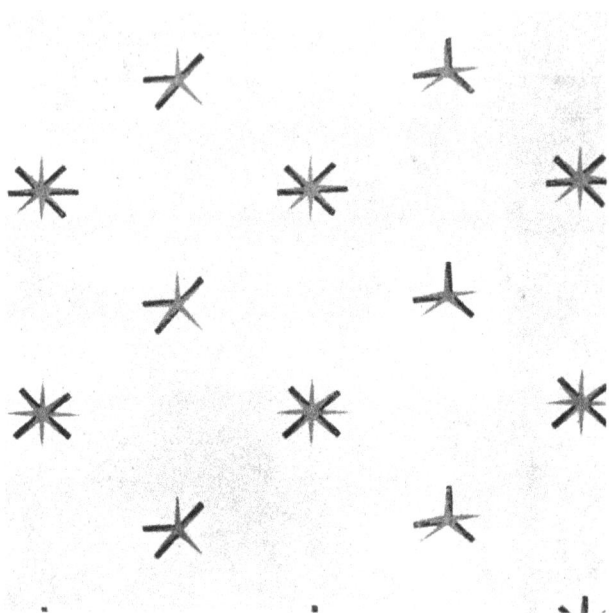

PLATE XIII.—*Decanter and glasses designed by Keith Murray and made by Stevens and Williams, Limited.*

PLATE XIV.—To the right is a composition in lime-yellow, dove-grey and silver lustre on white Foley China by Albert Rutherston, A.R.W.S.

A freehand design in old silver resist by Milner Gray, carried out in white Foley China. (See footnote on page 87.)

P L A T E X V.—A composition by Paul Nash in brown, earth-red, putty, and soft blue, carried out by freehand drawing on white Foley China.

A composition by Dame Laura Knight, A.R.A. The horse's head is engraved, and the plumes are carried out in freehand drawing in viridian green. The scallops forming the border are in old Staffordshire plum lustre. (Foley China.)

PLATE XVI.—Above: Three drinking mugs designed by Keith Murray for Josiah Wedgwood and Sons, Limited. To the left is a design by Dame Laura Knight, A.R.A., and below designs by Gordon Forsyth, all executed in Bizarre by Clarice Cliff and produced by A. J. Wilkinson, Limited, Royal Staffordshire Pottery, Burslem.

For Product Safety Concerns and Information please contact our EU
representative GPSR@taylorandfrancis.com
Taylor & Francis Verlag GmbH, Kaufingerstraße 24, 80331 München, Germany

www.ingramcontent.com/pod-product-compliance
Lightning Source LLC
Chambersburg PA
CBHW050440280326
41932CB00013BA/2187